W0051180

TOPICS IN MATHEMATICAL PHYSICS

PROBLEMY MATEMATICHESKOI FIZIKI

ПРОБЛЕМЫ МАТЕМАТИЧЕСКОЙ ФИЗИКИ

TOPICS IN MATHEMATICAL PHYSICS

Series editor: M. Sh. Birman

Volume 1: SPECTRAL THEORY AND WAVE PROCESSES—1967

The Lamb Problem for an Inhomogeneous Elastic Half-Space • A. G. Alenitsyn

Stieltjes Double-Integral Operators, I • M. Sh. Birman and M. Z. Solomyak

The Inverse Problem in the Theory of Seismic Wave Propagation • A. S. Blagoveshchenskii

The Trace Formulas and Some Asymptotic Estimates of the Resolvent Kernel of the Three-Dimensional Schroedinger Equation • V. S. Buslaev

The Nonself-Adjoint Schroedinger Operator, I • B. S. Pavlov

Volume 2: SPECTRAL THEORY AND PROBLEMS IN DIFFRACTION—1968

The Asymptotic Behavior of "Quasi-Eigenvalues" of the Exterior Problem for the Laplace Operator • V. M. Babich

Eigenfunctions Concentrated near a Closed Geodesic • V. M. Babich and V. F. Lazutkin

Stieltjes Double-Integral Operators, II • M. Sh. Birman and M. Z. Solomyak

Wave Propagation near the Curved Surface of an Inhomogeneous Body • V. S. Buldyrev

Continuum Integrals and the Asymptotic Behavior of the Solutions of Parabolic Equations as $t \to 0$. Applications to Diffraction • V. S. Buslaev

Eigenvalues of the Perturbed Schroedinger Operator with a Periodic Potential • V. A. Zheludev

Diffraction on a Convex Contour with Smoothly Varying Radius of Curvature and Impedance • I. A. Molotkov

The Nonself-Adjoint Schroedinger Operator, II • B. S. Pavlov

Volume 3: SPECTRAL THEORY—1969

The Asymptotic Behavior of the Solutions of the Wave Equation Concentrated Near the Axis of a Two-Dimensional Waveguide in an Inhomogeneous Medium • B. S. Buldyrev

Perturbations of the Spectrum of the Schroedinger Operator with a Complex Periodic Potential • V. A. Zheludev

The Discrete Spectra of the Dirac and Pauli Operators • O. I. Kurbenin

The Nonself-Adjoint Schroedinger Operator. III. • B. S. Pavlov

The Singular Numbers of the Sum of Completely Continuous Operators • S. Yu. Rotfel'd

Double-Integral Operators in the Ring $\hat{R}$ • M. Z. Solomyak

Correction to "The Inverse Problem in the Theory of Seismic Wave Propagation" • A. S. Blagoveshchenskii

TOPICS IN MATHEMATICAL PHYSICS
Volume 3

SPECTRAL THEORY

Edited by
M. Sh. Birman
Department of Physics
Leningrad State University

Translated from Russian

CONSULTANTS BUREAU · NEW YORK–LONDON · 1969

The original Russian text, published by Leningrad University Press in 1968, has been corrected by the editor for this edition.

ПРОБЛЕМЫ МАТЕМАТИЧЕСКОЙ ФИЗИКИ

Выпуск 3

Спектральная теория

ISBN 978-1-4684-7591-3 ISBN 978-1-4684-7589-0 (eBook)
DOI 10.1007/978-1-4684-7589-0

Library of Congress Catalog Card Number 78 - 93768

© 1969 Consultants Bureau, New York
A Division of Plenum Publishing Corporation
227 West 17th Street, New York, N. Y. 10011

United Kingdom edition published by Consultants Bureau, London
A Division of Plenum Publishing Company, Ltd.
Donington House, 30 Norfolk Street, London, W.C. 2, England

All rights reserved

No part of this publication may be reproduced in any
form without written permission from the publisher

CONTENTS

THE ASYMPTOTIC BEHAVIOR OF THE SOLUTIONS OF THE WAVE EQUATION CONCENTRATED NEAR THE AXIS OF A TWO-DIMENSIONAL WAVEGUIDE IN AN INHOMOGENEOUS MEDIUM

B. S. Buldyrev

The present article describes a method for the construction of the solutions of the wave equation

$$\frac{\partial^2 u}{\partial x^2} + \frac{\partial^2 u}{\partial z^2} + \frac{\omega^2}{c^2(x, z)} u = 0, \tag{1}$$

which describe as $\omega \to \infty$ the propagation of waves near the axis of a waveguide.

In the first section we give the conditions under which we can speak of a "waveguide-like" propagation of waves in an inhomogeneous medium and then we define the term "waveguide axis" used by us.

In the second section we construct the solutions of Eq. (1) which are concentrated near the waveguide axis and which decay exponentially outside a band which contains the waveguide axis. These solutions are then used (Section 3) to obtain asymptotic formulas for the corresponding subsequence of eigenfunctions and eigenvalues in the case of some boundary problems involving Eq. (1).

Babich and Lazutkin [1] have considered similar problems. With the help of the parabolic-equation method, they have studied the asymptotic behavior of the eigenfunctions of the Laplace operator that are concentrated in the vicinity of a closed geodesic on an arbitrary surface. Lazutkin [2] has studied the eigenfunctions of the Laplace operator concentrated in the vicinity of the stable diamaters of a closed two-dimensional region. He has also proposed a scheme for the construction of higher-order approximations by the parabolic-equation method. The procedure for the construction of higher-order approximations proposed in the present article is different from Lazutkin's procedure.

§ 1. A Waveguide in an Inhomogeneous Medium

Let L be a sufficiently smooth curve in an inhomogeneous medium in which the velocity of wave propagation is c(x, z). Let us establish the conditions under which the rays of geometrical optics, i.e., the extremums of the integral

$$I = \int \frac{dS}{c(x, z)}, \tag{1.1}$$

subtending small angles with the curve L over their whole length, will be situated in the vicinity of L. The complete solution of this problem reduces to the investigation of the stability of the solutions of dynamic systems and is quite complicated. We will restrict our investigation of stability to the first approximation, i.e., we will investigate the stability of the solutions of the linearized Euler equation. Let us assume that the curve L can be represented parametrically as

$$x = x(s), \quad z = z(s), \qquad -\infty < s < \infty,$$

where the parameter s is taken to be the arc length along L measured in a given direction from some initial point. Let us introduce a system of coordinates (s, n) in the vicinity of L on the (x, z) plane. The coordinates (s, n) of an arbitrary point M will be defined as follows: n is the distance along the normal drawn from M to L and s is the arc length along L from the initial point to the base of the normal. In addition, we will assume that n is positive if M is to the left of L as we move along L in the direction of increasing values of s.

Let us establish the relationship between the Cartesian coordinates (x, z) and the coordinates (s, n). It is obvious that we have

$$x = x(s) - n \sin\alpha,$$
$$z = z(s) + n \cos\alpha,$$

where α is the angle of inclination of the tangent to L. Since we have taken the parameter of the parametric representation to be the arc length s along L, we have $\cos\alpha = x'(s)$, $\sin\alpha = z'(s)$ and, consequently,

$$x = x(s) - z'(s)\, n,$$
$$z = z(s) + x'(s)\, n. \tag{1.2}$$

We will assume that the radius of curvature $\rho(s)$ of L is positive when the center of curvature is to the right of L and negative when the center of curvature is to the left of L. We have

$$\rho^{-1}(s) = z'(s)\, x''(s) - x'(s)\, z''(s). \tag{1.3}$$

Using expressions (1.2) and (1.3), we find that

$$dS^2 = dx^2 + dy^2 = \left(1 + \frac{n}{\rho(s)}\right)^2 ds^2 + dn^2.$$

Thus, the Lamé coefficients h_1 and h_2 of the coordinate system (s, n) are

$$h_1 = 1 + \frac{n}{\rho(s)}, \quad h_2 = 1. \tag{1.4}$$

Let ρ_{min} denote the minimum value of the absolute magnitude of the radius of curvature of L. It is clear that the coordinate system (s, n) is regular within the band $|n| < \rho_{min}$ surrounding L. As before, we will denote the velocity of wave propagation in the coordinates s, n by c(s, n) and we will assume that c(s, n) can be differentiated a sufficient number of times with respect to both s and n.

The fundamental functional of geometrical optics (1.1) expressed in terms of the coordinates s and n is

$$I=\int\frac{\sqrt{\left(1+\frac{n}{\rho(s)}\right)^2+\left(\frac{dn}{ds}\right)^2}}{c(s,n)}ds. \tag{1.5}$$

We will assume that the deviation of rays from L is small both in distance and in the angle of inclination of the tangent, i.e., we will assume that $|n|$ and $|dn/ds|$ are small.* Let us expand the integrand

$$\Phi\left(s,n,\frac{dn}{ds}\right)=\frac{\sqrt{\left(1+\frac{n}{\rho(s)}\right)^2+\left(\frac{dn}{ds}\right)^2}}{c(s,n)}$$

in a series in powers of n and dn/ds neglecting cubic and smaller terms

$$\Phi\left(s,n,\frac{dn}{ds}\right)=\frac{1}{c_0(s)}\left\{1+\left[\frac{1}{\rho(s)}-\frac{c_1(s)}{c_0(s)}\right]n+\left[\frac{c_1^2(s)}{c_0^2(s)}-\right.\right.$$
$$\left.\left.-\frac{1}{2}\frac{c_2(s)}{c_0(s)}-\frac{1}{\rho(s)}\frac{c_1(s)}{c_0(s)}\right]n^2+\frac{1}{2}\left(\frac{dn}{ds}\right)^2+O\left[n^3,\ n\left(\frac{dn}{ds}\right)^2\right]\right\},$$

where $c_0(s)=c(s,0)$, $c_1(s)=\left.\frac{\partial c(s,n)}{\partial n}\right|_{n=0}$, $c_2(s)=\left.\frac{\partial^2 c(s,n)}{\partial n^2}\right|_{n=0}$. This expansion allows us to write the Euler equation

$$\Phi_n-\frac{d}{ds}\Phi_{n'}=0$$

as

$$\frac{d^2n}{ds^2}-\frac{c_0'(s)}{c_0(s)}\frac{dn}{ds}-\left[2\left(\frac{c_1(s)}{c_0(s)}\right)^2-\frac{2}{\rho(s)}\frac{c_1(s)}{c_0(s)}-\frac{c_2(s)}{c_0(s)}\right]n-\left[\frac{1}{\rho(s)}-\frac{c_1(s)}{c_0(s)}\right]+(\text{quadratic terms})=0. \tag{1.6}$$

It follows from Eq. (1.6) that the curve n = 0, i.e., L itself is an extremum of the functional (1.5) or a geometrical ray if and only if the equality

$$\frac{1}{\rho(s)}=\frac{c_1(s)}{c_0(s)} \tag{1.7}$$

holds on L. In the following, we assume that L is a ray of geometric optics and, consequently, that equality (7) is always satisfied. With the help of equality (7) we can rewrite (1.6) as

$$\frac{d^2n}{ds^2}-\frac{c_0'(s)}{c_0(s)}\cdot\frac{dn}{ds}+\frac{c_{n^2}''(s)}{c_0(s)}+(\text{quadratic terms})=0.$$

* More accurately, we assume that the dimensionless quantities n/ρ and $\left.\frac{1}{c}\frac{\partial^k c}{\partial n^k}\right|_{n=0}n^k$ are small.

Assuming that

$$n(s)=\frac{c_0^{1/2}(s)}{c_0^{1/2}(0)}\,y(s),$$

we obtain an equation without a first derivative

$$\frac{d^2y}{ds^2}+K(s)\,y+(\text{ quadratic terms })=0,$$

in which

$$K(s)=\frac{1}{2}\frac{c_0''(s)}{c_0(s)}-\frac{3}{4}\left[\frac{c_0'(s)}{c_0(s)}\right]^2+\frac{c_2(s)}{c_0(s)}\,.$$

Thus, in the first approximation the rays that are close to L are described by the equation

$$\frac{d^2y}{ds^2}+K(s)\,y=0. \tag{1.8}$$

When $K(s) \leq 0$, the solutions of Eq. (1.8) that are not identically zero become zero at most once (the ray intersects L at most once). In this case, at least one of the two linearly independent solutions of Eq. (1.8) grows unboundedly as $s \to \infty$.

On the other hand, if $K(s) \geq m > 0$, the solutions of Eq. (1.8) have an infinite number of zeros. In any segment of length greater than $\pi/\sqrt{m}$, any solution of Eq. (1.8) will become equal to zero at least once (Sturm's theorem). The rays will "twist" around L and as $s \to \infty$ they will intersect L an infinite number of times. It can be shown [3] that if the function K(s) increases monotonically for $s > s_1$ and tends to a constant limit $M > 0$, then the solutions of Eq. (1.8) remain bounded as $s \to \infty$. Let the initial conditions

$$y(0)=0 \text{ and } y'(0)=\alpha$$

hold at the point $s = 0$, then for a sufficiently small α the absolute magnitude of the function y(s) will not exceed a given quantity over the whole of the infinte interval $0 \leq s < \infty$. In this case, we will say that the ray system constructed in the vicinity of L is stable in the first approximation. The condition governing the behavior of K(s) at infinity will be assumed always to hold. In most cases, however, we will only be interested in the behavior of rays over a finite segment of L, so that the condition governing the behavior of K(s) at infinity will not be relevant to the discussion.

In the following, the finite segment of L, i.e., an arc l $(0 \leq s \leq s_0)$, will be called the waveguide axis and we will speak of a waveguide mode of wave propagation along l if l is a geometrical ray (an extremum of integral (1.5)) and if the rays repeatedly intersect l, i.e., if the following condition holds

$$K(s)>0. \tag{1.9}$$

In the special case when L is a straight line $[\rho(s) = \infty]$ and the velocity is independent of s [i.e., when $c(s, n) = c(n)$], these conditions reduce to

$$c'_n(0)=0, \quad c''_{n^2}(0)>0,$$

i.e., the requirement that the velocity is a minimum along the ray being considered. It is easy to show that in this case the ray system in the vicinity of the straight line L will be stable "in the larger."

§ 2. The Construction of the Solutions of the Wave Equation Concentrated Near the Waveguide Axis

Let L be a curve in the (x, z) plane and let conditions (1.7) and (1.9) be satisfied along L, i.e., L is the waveguide axis. Let us now construct solutions of the wave equation (1) which are concentrated near L and which decrease exponentially outside a band containing L.

1. First of all, let us consider the special case when L coincides with the OX axis and

$$c(x, z)=c_0\left[1-\frac{1}{4}\left(\frac{z}{z_0}\right)^2\right]^{-1/2},$$

where c_0 and z_0 are constants. In this case, the variables in Eq. (1) can be separated and the exact solution

$$u_q(x, z)=\exp\left[\pm i\frac{\omega}{c_0}\sqrt{1-\frac{c_0}{z_0}\frac{q+1/2}{\omega}}\,x\right]D_q\left(\frac{\omega^{1/2}}{\sqrt{c_0 z_0}}z\right). \tag{2.1}$$

Here $D_q(\nu)$, with $\nu = \frac{\omega^{1/2}}{\sqrt{c_0 z_0}}\cdot z$, are parabolic cylinder functions; q is a constant which arises in the separation of variables in Eq. (1.1). It is known that the parabolic cylinder function $D_q(\nu)$ tends to zero as $\nu \to \pm\infty$ only when q is a positive integer or zero. Therefore, in the following we will consider that q in formula (2.1) is a positive integer or zero since we are interested in solutions of Eq. (1) tending to zero as $z \to \pm\infty$. With integer q, the functions $D_q\left(\frac{\omega^{1/2}}{\sqrt{c_0 z_0}}z\right)$ oscillate when $\left|\frac{\omega^{1/2}}{\sqrt{c_0 z_0}}z\right| < 2\sqrt{q+1/2}$, and decrease exponentially when $\left|\frac{\omega^{1/2}}{\sqrt{c_0 z_0}}z\right| > 2\sqrt{q+1/2}$. Thus, the solutions given by (2.2) for large ω are appreciably different from zero only in the narrow band

$$|z| \leqslant 2\sqrt{c_0 z_0}(q+1/2)^{1/2}\omega^{-1/2}$$

around the OX axis and are exponentially small outside this band. When $\omega < (q + 1/2)\, c_0/z_0$, the solution $u_q(x, z)$ will increase or decrease exponentially with increasing x. When $\omega > (q + 1/2)\, c_0/z_0$, we obtain oscillating solutions or the so-called waveguide solutions of Eq. (1). When $\omega \gg (q + 1/2)\, c_0/z_0$, the square root in formula (2.1) can be expanded in a series and we obtain

$$u_q(x, z)=\exp\left\{\pm i\frac{\omega}{c_0}\left[x-\frac{1}{2}\frac{c_0}{z_0}\frac{q+1/2}{\omega}x-\frac{1}{8}\left(\frac{c_0}{z_0}\frac{q+1/2}{\omega}\right)^2 x+\ldots\right]\right\}D_q\left(\frac{\omega^{1/2}}{\sqrt{c_0 z_0}}z\right).$$

2. Let us now consider the general case. Let L be a curve in the (x, z) plane and such that conditions (1.7) and (1.9) are satisfied along L. Let us rewrite the wave equation (1) in terms of the variables s and n associated with L. Making use of formulas (1.4) for the Lamé coefficients, we obtain

$$\frac{\partial}{\partial s}\left[\left(1+\frac{n}{\rho(s)}\right)^{-1}\frac{\partial u}{\partial s}\right]+\frac{\partial}{\partial n}\left[\left(1+\frac{n}{\rho(s)}\right)\frac{\partial u}{\partial n}\right]+\frac{\omega^2}{c^2(s, n)}\left(1+\frac{n}{\rho(s)}\right)u=0. \tag{2.2}$$

We will assume that the wave-propagation velocity c(s, n) in Eq. (2.2) can be represented by a Taylor series expansion in n

$$c(s, n) = c_0(s) + c_1(s)\,n + \frac{1}{2}c_2(s)\,n^2 + \ldots + \frac{1}{(k-1)!}c_{k-1}(s)\,n^{k-1} + O(n^k). \tag{2.3}$$

Since we are only interested in the waveguide solutions of Eq. (2.2) for sufficiently large values of ω, i.e., solutions which by analogy with the separable case must be concentrated in the vicinity of L in a band whose width is of order $\omega^{-1/2}$, let us perform a change of variable in Eq. (2.2) and introduce the new variable

$$\nu = \omega^{1/2} n. \tag{2.4}$$

In order to have ν a dimensionless variable, we must introduce a constant factor $(c_0 z_0)^{-1/2}$ characterizing the unit of velocity and distance measurement in formula (2.4). However, we will omit this factor right from the start, assuming that it is equal to unity, so as not to complictate the formulas. The wave field u(s, n) in the variables s and ν will be denoted as before by u(s, ν).

In the following we will see that the function u(s, ν) and its partial derivatives $\partial^m u/\partial\nu^m$ will be of the same order of magnitude with respect to the large parameter ω when $|\nu| <$ const. The equation for u(s, ν) in terms of the variables s and ν is

$$\omega A\frac{\partial^2 u}{\partial\nu^2} + \omega^{\frac{1}{2}} B\frac{\partial u}{\partial\nu} + C\frac{\partial^2 u}{\partial s^2} + \omega^{-\frac{1}{2}} D\frac{\partial u}{\partial s} + \omega^2 E u = 0, \tag{2.5}$$

where

$$A = 1 + \omega^{-\frac{1}{2}}\frac{\nu}{\rho(s)},\quad B = \frac{1}{\rho(s)},\quad C = \left(1 + \omega^{-\frac{1}{2}}\frac{\nu}{\rho(s)}\right)^{-1},$$

$$D = \frac{\nu}{\rho^2(s)}\rho'(s)\left(1 + \omega^{-\frac{1}{2}}\frac{\nu}{\rho(s)}\right)^{-2},$$

$$E = \left(1 + \omega^{-\frac{1}{2}}\frac{\nu}{\rho(s)}\right)c^{-2}\left(s,\ \nu\omega^{-\frac{1}{2}}\right).$$

Let us represent C, D, and E as series in powers of $\omega^{-1/2}$ with the help of expansion (2.3)

$$C = \sum_{m=0}^{N-1}(-1)^m\left(\frac{\nu}{\rho(s)}\right)^m \omega^{-\frac{m}{2}} + O\left(\nu^N \omega^{-\frac{N}{2}}\right), \tag{2.6}$$

$$D = \frac{\nu}{\rho^2(s)}\cdot\rho'(s)\left\{\sum_{m=0}^{N-1}(-1)^m(m+1)\left(\frac{\nu}{\rho(s)}\right)^m \omega^{-\frac{m}{2}} + O\left(\nu^N \omega^{-\frac{N}{2}}\right)\right\}, \tag{2.7}$$

$$E = \frac{1}{c_0^2(s)} + \sum_{m=1}^{N-1} e_m(s)\,\nu^m \omega^{-\frac{m}{2}} + O\left(\nu^N \omega^{-\frac{N}{2}}\right), \tag{2.8}$$

where

$$e_1(s)=\frac{1}{c_0^2(s)}\left[\frac{1}{\rho(s)}-2\frac{c_1(s)}{c_0(s)}\right], \tag{2.9}$$

$$e_2(s)=-\frac{1}{c_0^2(s)}\left[2\frac{c_1(s)}{c_0(s)}\frac{1}{\rho(s)}+\frac{c_2(s)}{c_0(s)}-3\frac{c_1^2(s)}{c_0^2(s)}\right], \tag{2.10}$$

$$e_3(s)=-\frac{1}{c_0^2(s)}\left[\left(\frac{c_2(s)}{c_0(s)}-3\frac{c_1^2(s)}{c_0^2(s)}\right)\frac{1}{\rho(s)}+\frac{1}{3}\frac{c_3(s)}{c_0(s)}-3\frac{c_1(s)c_2(s)}{c_0^2(s)}+4\frac{c_1^3(s)}{c_0^3(s)}\right], \tag{2.11}$$

and so on. The coefficients of $\omega^{-1/2}$ in expansions (2.6)-(2.8) are polynomials in ν whose coefficients are in turn functions of s. Therefore, by analogy with the separable case, we will seek solutions of Eq. (2.5) in the form

$$u(s,\nu)=\exp\{i\omega\Phi(s,\nu;\omega)\}D_q\{\Psi(s,\nu;\omega)\}. \tag{2.12}$$

Here, $D_q(\Psi)$, $q = 0, 1, 2, \ldots$ are parabolic cylinder functions and $\Phi(s, \nu; \omega)$ and $\Psi(s, \nu; \omega)$ are given by

$$\Phi(s,\nu;\omega)=\alpha_0(s,\nu)+\alpha_1(s,\nu)\omega^{-\frac{1}{2}}+\alpha_2(s,\nu)\omega^{-1}+\ldots, \tag{2.13}$$

$$\Psi(s,\nu;\omega)=\beta_0(s,\nu)+\beta_1(s,\nu)\omega^{-\frac{1}{2}}+\beta_2(s,\nu)\omega^{-1}+\ldots, \tag{2.14}$$

where $\alpha_n(s, \nu)$ and $\beta_n(s, \nu)$, $n = 0, 1, \ldots$, as in expansions (2.6)-(2.8), will be assumed to be polynomials in s, i.e.,

$$\alpha_n(s,\nu)=\sum_{k=0}^{p}\alpha_{nk}(s)\nu^k, \tag{2.15}$$

$$\beta_n(s,\nu)=\sum_{k=0}^{q}\beta_{nk}(s)\nu^k. \tag{2.16}$$

The aim of our subsequent calculations is to find equations that must be satisfied by $\alpha_n(s, \nu)$ and $\beta_n(s, \nu)$ so that $u(s, \nu)$ defined by (2.12) formally satisfies Eq. (2.5). After $u(s, \nu)$ is substituted into Eq. (2.5), the latter will contain the parabolic cylinder functions $D_q(\Psi)$ and their derivatives $D_q'(\Psi)$ and $D_q''(\Psi)$.

With the help of the equation $D_q''(\Psi) + (q + 1/2 - \Psi^2/4)\,D_q(\Psi) = 0$ satisfied by the parabolic cylinder functions, we can express $D_q''(\Psi)$ in terms of $D_q(\Psi)$, so that Eq. (2.5) after the substitution of $u(s, \nu)$ and the cancellation of the exponential factor becomes

$$a(s,\nu;\omega)D_q\{\Psi(s,\nu;\omega)\}+b(s,\nu;\omega)D_q'\{\Psi(s,\nu;\omega)\}=0, \tag{2.17}$$

where

$$a(s,\nu;\omega)=-A\left(\frac{\partial\Phi}{\partial\nu}\right)^2\omega^3+\left[iA\frac{\partial^2\Phi}{\partial\nu^2}-C\left(\frac{\partial\Phi}{\partial s}\right)^2+E\right]\omega^2+iB\frac{\partial\Phi}{\partial\nu}\omega^{3/2}+$$

$$+\left[-A\left(\frac{\partial\Psi}{\partial\nu}\right)^2\left(q+1/2-\frac{1}{4}\Psi^2\right)+iC\frac{\partial^2\Phi}{\partial s^2}\right]\omega^1+\ iD\frac{\partial\Phi}{\partial s}\,\omega^{1/2}-C\left(\frac{\partial\Psi}{\partial s}\right)^2\left(q+\frac{1}{2}-\frac{1}{4}\Psi^2\right), \tag{2.18}$$

$$b(s,\ \nu;\ \omega)=2iA\frac{\partial\Phi}{\partial\nu}\frac{\partial\Psi}{\partial\nu}\omega^2+\left[A\frac{\partial^2\Psi}{\partial\nu^2}+2iC\frac{\partial\Phi}{\partial s}\frac{\partial\Psi}{\partial s}\right]\omega^1\ +C\frac{\partial^2\Psi}{\partial s^2}+D\frac{\partial\Psi}{\partial s}\,\omega^{-\frac{1}{2}}. \tag{2.19}$$

Equation (2.17) is satisfied only when the coefficients of $D_q(\Psi)$ and $D_q'(\Psi)$ are both equal to zero, i.e., when

$$a(s,\ \nu;\ \omega)=0,\ b(s,\ \nu;\ \omega)=0; \tag{2.20}$$

equalities (2.20) must be satisfied identically with respect to ω. Substituting expansions (2.13), (2.14), and (2.6)-(2.8) into formulas (2.18) and (2.19) and collecting coefficients of like powers of $\omega^{-1/2}$, we obtain

$$a(s,\ \nu;\ \omega)=\omega^2\sum_{m=-2}^{\infty}a_m(s,\ \nu)\,\omega^{-\frac{m}{2}},$$

$$b(s,\ \nu;\ \omega)=\omega^2\sum_{m=0}^{\infty}b_m(s,\ \nu)\,\omega^{-\frac{m}{2}}.$$

The coefficients $a_m(s, \nu)$ and $b_m(s, \nu)$ here contain partial derivatives of $\alpha_n(s, \nu)$ and $\beta_n(s, \nu)$ with respect to s and ν. Equations (2.20) will hold identically with respect to ω if and only if

$$\begin{aligned}&a_m(s,\ \nu)=0,\ m=-2,\ -1,\ 0,\ 1\ldots,\\&b_m(s,\ \nu)=0,\ m=0,\ 1,\ \ldots.\end{aligned} \tag{2.21}$$

Equalities (2.21) represent a system of recurrence relations involving partial derivatives for the polynomials $\alpha_n(s, \nu)$ and $\beta_n(s, \nu)$.

Equations (2.21) are written out in expanded form below and their solutions are then constructed.

3. Equations (2.21) for large m can be simplified if we first obtain the solutions of the first few equations of system (2.21). After the necessary calculations, we obtain

$$a_{-2}(s,\ \nu)\equiv-\left(\frac{\partial\alpha_0}{\partial\nu}\right)^2=0, \tag{2.22}$$

$$a_{-1}(s,\ \nu)\equiv-2\frac{\partial\alpha_0}{\partial\nu}\cdot\frac{\partial\alpha_1}{\partial\nu}-\frac{\partial\alpha_0}{\partial\nu}\cdot\frac{\nu}{\rho}=0, \tag{2.23}$$

$$a_0(s,\ \nu)\equiv-\left(\frac{\partial\alpha_1}{\partial\nu}\right)^2-2\frac{\partial\alpha_0}{\partial\nu}\frac{\partial\alpha_2}{\partial\nu}-2\frac{\partial\alpha_0}{\partial\nu}\frac{\partial\alpha_1}{\partial\nu}\frac{\nu}{\rho}+i\frac{\partial^2\alpha_0}{\partial\nu^2}-\left(\frac{\partial\alpha_0}{\partial s}\right)^2+\frac{1}{c_0^2(s)}=0, \tag{2.24}$$

$$a_1(s, \nu) \equiv -2\frac{\partial\alpha_1}{\partial\nu}\frac{\partial\alpha_2}{\partial\nu} - \left(\frac{\partial\alpha_1}{\partial\nu}\right)^2\frac{\nu}{\rho} + i\frac{\partial^2\alpha_1}{\partial\nu^2} - 2\frac{\partial\alpha_0}{\partial s}\frac{\partial\alpha_1}{\partial s} + \left(\frac{\partial\alpha_0}{\partial s}\right)^2\frac{\nu}{\rho} + e_1(s)\,\nu = 0, \tag{2.25}$$

$$b_0(s, \nu) \equiv 2i\frac{\partial\alpha_0}{\partial\nu}\frac{\partial\beta_0}{\partial\nu} = 0, \tag{2.26}$$

$$b_1(s, \nu) \equiv 2i\frac{\partial\alpha_1}{\partial\alpha}\frac{\partial\beta_0}{\partial\nu} + 2i\frac{\partial\alpha_0}{\partial\nu}\frac{\partial\beta_1}{\partial\nu} + 2i\frac{\partial\alpha_0}{\partial\nu}\frac{\partial\beta_0}{\partial\nu}\frac{\nu}{\rho} = 0. \tag{2.27}$$

Equation (2.22) shows that $\alpha_0(s, \nu)$ is independent of ν and, consequently, is a polynomial of degree zero in ν, i.e.,

$$\alpha_0(s, \nu) = \alpha_{00}(s). \tag{2.28}$$

In view of (2.28), Eqs. (2.23) and (2.26) are automatically satisfied, while Equation (2.27) reduces to

$$2i\frac{\partial\alpha_1}{\partial\nu}\frac{\partial\beta_0}{\partial\nu} = 0.$$

This equation will be satisfied both when $\partial\alpha_1/\partial\nu = 0$ and when $\partial\beta_0/\partial\nu = 0$. However, if we assume that $\partial\beta_0/\partial\nu = 0$, we would contradict the special case (2.1) of separability when $\partial\beta_0/\partial\nu \neq 0$. Assuming that $\partial\alpha_1/\partial\nu = 0$, we come to the conclusion that

$$\alpha_1(s, \nu) = \alpha_{10}(s). \tag{2.29}$$

Substituting (2.28) and (2.29) into Eq. (2.24), we find that

$$\left(\frac{d\alpha_{00}}{ds}\right)^2 = \frac{1}{c_0^2(s)},$$

from which we obtain

$$\alpha_{00}(s) = \pm\int\limits_{C_0}^{s}\frac{d\tau}{c_0(\tau)}, \tag{2.30}$$

where C_0 is an arbitrary constant. The positive sign is taken in formula (2.30) for all subsequent derivations.

Since we have $\partial\alpha_1/\partial\nu = 0$, $\partial\alpha_0/\partial s = 1/c_0(s)$, the coefficient $e_1(s)$ is defined by formula (2.9), and relation (1.7) is satisfied, namely,

$$\frac{1}{\rho(s)} = \frac{c_1(s)}{c_0(s)}, \tag{2.31}$$

we see that Eq. (2.25) yields

$$\frac{\partial\alpha_1}{\partial s} = 0, \tag{2.32}$$

from which we obtain

$$\alpha_1(s, \nu) = \alpha_{10}(s) = C_1,$$

where C_1 is an arbitrary constant.

4. The subsequent equations of system (2.21) will be written down with the previously established relations $\partial\alpha_0/\partial\nu = \partial\alpha_1/\partial\nu = \partial\alpha_1/\partial s = 0$ and $\partial\alpha_0/\partial s = 1/c_0(s)$ taken into account. For m = 2, we have

$$a_2(s,\nu)\equiv-\left(\frac{\partial\alpha_2}{\partial\nu}\right)^2+i\frac{\partial^2\alpha_2}{\partial\nu^2}-2\frac{1}{c_0(s)}\frac{\partial\alpha_2}{\partial s}-\frac{1}{c_0^2(s)}\frac{\nu^2}{\rho^2}+$$

$$+e_2(s)\nu^2-\left(\frac{\partial\beta_0}{\partial\nu}\right)^2\left(q+\frac{1}{2}-\frac{1}{4}\beta_0^2\right)-i\frac{c_0'(s)}{c_0^2(s)}=0, \tag{2.33}$$

$$b_2(s,\nu)\equiv 2i\frac{\partial\alpha_2}{\partial\nu}\frac{\partial\beta_0}{\partial\nu}+\frac{\partial^2\beta_0}{\partial\nu^2}+2i\frac{1}{c_0(s)}\frac{\partial\beta_0}{\partial s}=0. \tag{2.34}$$

On the basis of formulas (2.10) and (2.31), we can write the coefficient of ν^2 in Eq. (2.33) as

$$-\frac{1}{c_0^2(s)}\frac{1}{\rho^2(s)}+e_2(s)=-\frac{c_2(s)}{c_0^3(s)}.$$

Equations (2.33) and (2.34) define the polynomials $\alpha_2(s,\nu)$ and $\beta_0(s,\nu)$, It follows from Eq. (2.34) that $\alpha_2(s,\nu)$ is a second-degree polynomial

$$\alpha_2(s,\nu)=\alpha_{00}(s)+\alpha_{21}(s)\nu+\alpha_{22}(s)\nu^2. \tag{2.35}$$

If $\alpha_2(s,\nu)$ is a second-degree polynomial, then Eq. (2.33) shows that $\beta_0(s,\nu)$ is a first-degree polynomial

$$\beta_0(s,\nu)=\beta_{00}(s)+\beta_{01}(s)\nu. \tag{2.36}$$

Let us substitute polynomials (2.35) and (2.36) into Eqs. (2.33) and (2.34) and let us collect together the coefficients of like powers of ν; we obtain

$$\left(-4\alpha_{22}^2-2\frac{1}{c_0}\alpha_{22}'-\frac{c_2}{c_0^3}+\frac{1}{4}\beta_{01}^4\right)\nu^2+\left(-4\alpha_{22}\alpha_{21}-2\frac{1}{c_0}\alpha_{21}'+\frac{1}{2}\beta_{01}^3\beta_{00}\right)\nu+$$

$$+\left(-\alpha_{21}^2+2i\alpha_{22}-2\frac{1}{c_0}\alpha_{20}'-\left(q+\frac{1}{2}\right)\beta_{01}^2+\frac{1}{4}\beta_{01}^2\beta_{00}^2-i\frac{c_0'}{c_0^2}\right)=0, \tag{2.37}$$

$$\left(2\alpha_{22}\beta_{01}+\frac{1}{c_0}\beta_{01}'\right)\nu+\left(\alpha_{21}\beta_{01}+\frac{1}{c_0}\beta_{00}'\right)=0. \tag{2.38}$$

These two equalities will be identically zero whatever the value of ν only if all the expressions in round brackets are zero. From (2.38) we find that

$$\alpha_{22}=-\frac{1}{2c_0}\frac{\beta_{01}'}{\beta_{01}},\quad \alpha_{21}=-\frac{1}{c_0}\frac{\beta_{00}'}{\beta_{01}} \tag{2.39}$$

Substituting expressions (2.39) into Eq. (2.37), and equating the coefficients of ν^2, ν^1, ν^0 to zero, we find equations for β_{01}, β_{00}, and α_{20}, namely,

$$\frac{1}{c_0}\left(\frac{1}{c_0}\frac{\beta_{01}'}{\beta_{01}}\right)'-\left(\frac{1}{c_0}\frac{\beta_{01}'}{\beta_{01}}\right)^2+\frac{1}{4}\beta_{01}^4-\frac{c_2}{c_0^3}=0; \tag{2.40}$$

$$2\frac{1}{c_0}\left(\frac{1}{c_0}\frac{\beta'_{00}}{\beta_{01}}\right)' - 2\frac{1}{c_0^2}\frac{\beta'_{01}}{\beta_{01}^2}\beta'_{00} + \frac{1}{2}\beta_{01}^3\beta_{00} = 0, \tag{2.41}$$

$$2\frac{1}{c_0}\alpha'_{20} = -i\frac{c'_0}{c_0^2} - i\frac{1}{c_0}\frac{\beta'_{01}}{\beta_{01}} - \left(q+\frac{1}{2}\right)\beta_{01}^2 - \frac{1}{c_0^2}\left(\frac{\beta'_{00}}{\beta_{01}}\right)^2 + \frac{1}{4}\beta_{01}^2\cdot\beta_{00}^2. \tag{2.42}$$

First of all, let us construct the solution of Eq. (2.40). Le us introduce a new unknown function in Eq. (2.40) with the help of

$$F(s) = c_0^{-1/2}(s)\cdot\beta_{01}^{-1}(s).$$

The equation satisfied by F(s) is

$$\frac{F''}{F} - \frac{1}{4}\frac{1}{F^4} + \frac{1}{2}\frac{c''_0(s)}{c_0(s)} - \frac{3}{4}\left(\frac{c'_0(s)}{c_0(s)}\right)^2 + \frac{c_2(s)}{c_0(s)} = 0$$

or

$$F'' + K(s)F = \frac{1}{4}\frac{1}{F^3}, \tag{2.43}$$

where

$$K(s) = \frac{1}{2}\frac{c''_0(s)}{c_0(s)} - \frac{3}{4}\left(\frac{c'_0(s)}{c_0(s)}\right)^2 + \frac{c_2(s)}{c_0(s)}.$$

The coefficient K(s) in Eq. (2.43) is exactly equal to the coefficient K(s) in the linearized Euler equation (1.8) describing in the first approximation the rays that are close to L. Equation (2.43) has also been derived by Babich and Lazutkin [1] in their investigation of solutions of the wave equation concentrated near a closed geodesic. Let us take $\beta_{00}(s) = \beta_{01}(s)c^{1/2}(s)y(s)$ in Eq. (2.41). We find that the equation satisfied by y(s) is

$$y'' + \left(-\frac{F''}{F} + \frac{1}{4}\frac{1}{F^4}\right)y = 0.$$

Since F(s) satisfies (2.43), the above equation can be rewritten as

$$y'' + K(s)y = 0 \tag{2.44}$$

and, therefore, it coincides with the linearized Euler equation (1.8) mentioned above.

Let $y_1(s)$ and $y_2(s)$ be two linearly independent solutions of Eq. (2.44). Then, y(s) in the general case is given by

$$y(s) = a_1y_1(s) + a_2y_2(s),$$

where a_1 and a_2 are arbitrary constants.

The function F(s) is a solution of the nonlinear equation (2.43). However, it can also be expressed in terms of $y_1(s)$ and $y_2(s)$. Let $W[y_1, y_2]$ denote the Wronskian of the linearly independent solutions $y_1(s)$ and $y_2(s)$. Let $\|a\|$ be a symmetric second-order matrix such that

$$\det\|a_{k,j}\| W^2[y_1y_2] = \frac{1}{4}. \tag{2.45}$$

Then, F(s) can be expressed as [1] (see also [4])

$$F(s) = \left[\sum_{k,j=1}^{2} a_{k,j} y_k(s) y_j(s)\right]^{1/2}. \tag{2.46}$$

It should be noted that the right-hand side of (2.46) contains two arbitrary constants since the elements $a_{k,j}$ are subject to the two conditions $a_{12} = a_{21}$ and (2.45).

If the solutions of Eqs. (2.43) and (2.44) have been found and the functions F(s) and y(s) have been determined, then the coefficients $\alpha_{22}(s)$ and $\alpha_{21}(s)$ of the polynomial $\alpha_2(s, \nu)$ are given by formulas (2.39), while Eq. (2.42) for $\alpha_{20}(s)$ becomes

$$\alpha_{20}'(s) = i\left(-\frac{1}{4}\frac{c_0'}{c_0} + \frac{1}{2}\frac{F'}{F}\right) - \frac{q+\frac{1}{2}}{2}\frac{1}{F^2} - \frac{1}{2}\left(y' - \frac{F'}{F}y\right) + \frac{1}{8}\frac{y^2}{F^4},$$

from which we have

$$\alpha_{20}(s) = -\frac{i}{2}\ln c_0^{1/2}F^{-1} - \frac{q+\frac{1}{2}}{2}\int_{C_2}^{s}\frac{d\tau}{F^2(\tau)} + \frac{1}{2}\int_{C_2}^{s}\left[\frac{F'(\tau)}{F(\tau)} - \frac{y'(\tau)}{y(\tau)} + \frac{1}{4}\frac{y(\tau)}{F^4(\tau)}\right]y(\tau)\,d\tau,$$

where C_2 is an arbitrary constant.

5. Having constructed the polynomials $\alpha_2(s, \nu)$ and $\beta_0(s, \nu)$, we proceed to the determination of the next pair of polynomials $\alpha_3(s, \nu)$ and $\beta_1(s, \nu)$. Taking m = 3 in system (2.21), we obtain the following two equations:

$$\begin{aligned} a_3(s, \nu) = &-2\frac{\partial\alpha_2}{\partial\nu}\frac{\partial\alpha_3}{\partial\nu} + i\frac{\partial^2\alpha_3}{\partial\nu^2} - 2\frac{1}{c_0(s)}\frac{\partial\alpha_3}{\partial s} - 2\frac{\partial\beta_0}{\partial\nu}\frac{\partial\beta_1}{\partial\nu}\times \\ &\times\left(q+\frac{1}{2}-\frac{1}{4}\beta_0^2\right) + \frac{1}{2}\beta_0\beta_1\left(\frac{\partial\beta_0}{\partial\nu}\right)^2 + \left[\frac{1}{c_0^2(s)} - \frac{1}{\rho^3(s)} + e_3(s)\right]\nu^3 + \\ &+\left[-\left(\frac{\partial\alpha_2}{\partial\nu}\right)^2 + i\frac{\partial^2\alpha_2}{\partial\nu^2} + 2\frac{1}{c_0(s)}\frac{\partial\alpha_2}{\partial s} - \left(\frac{\partial\beta_0}{\partial\nu}\right)^2\left(q+\frac{1}{2}-\frac{1}{4}\beta_0^2\right) + \right. \\ &\left. + i\frac{c_0'(s)}{c_0^2(s)} + i\frac{1}{c_0(s)}\frac{\rho'(s)}{\rho(s)}\right]\frac{1}{\rho(s)}\nu + i\frac{\partial\alpha_2}{\partial\nu}\frac{1}{\rho(s)} = 0, \end{aligned} \tag{2.47}$$

$$\begin{aligned} b_3(s, \nu) = &\,2i\frac{\partial\alpha_2}{\partial\nu}\frac{\partial\beta_1}{\partial\nu} + 2i\frac{\partial\alpha_3}{\partial\nu}\frac{\partial\beta_0}{\partial\nu} + 2i\frac{1}{c_0(s)}\frac{\partial\beta_1}{\partial s} + \frac{\partial^2\beta_1}{\partial\nu^2} + \\ &+\left[2i\frac{\partial\alpha_2}{\partial\nu}\frac{\partial\beta_0}{\partial\nu} + \frac{\partial^2\beta_0}{\partial\nu^2} - 2i\frac{1}{c(s)}\frac{\partial\beta_0}{\partial s}\right]\frac{1}{\rho(s)}\nu + \frac{\partial\beta_0}{\partial\nu}\frac{1}{\rho(s)} = 0. \end{aligned} \tag{2.48}$$

Leaving the unknowns $\alpha_3(s, \nu)$ and $\beta_1(s, \nu)$ on the left-hand sides of the above equations and transferring all other terms to the right-hand sides, we can rewrite Eqs. (2.47) and (2.48) as

$$-2\frac{\partial\alpha_2}{\partial\nu}\frac{\partial\alpha_3}{\partial\nu} + i\frac{\partial^2\alpha_3}{\partial\nu^2} - 2\frac{1}{c_0(s)}\frac{\partial\alpha_3}{\partial s} - 2\frac{\partial\beta_0}{\partial\nu}\frac{\partial\beta_1}{\partial\nu}\left(q+\frac{1}{2}-\frac{1}{4}\beta_0^2\right) + \frac{1}{2}\beta_0\beta_1\left(\frac{\partial\beta_0}{\partial\nu}\right)^2 = \delta_3(s)\nu^3 + \ldots\delta_0(s), \tag{2.49}$$

$$2i\frac{\partial\alpha_3}{\partial\nu}\frac{\partial\beta_0}{\partial\nu} + 2i\frac{\partial\alpha_2}{\partial\nu}\frac{\partial\beta_1}{\partial\nu} + 2i\frac{1}{c_0(s)}\frac{\partial\beta_1}{\partial s} + \frac{\partial^2\beta_1}{\partial\nu^2} = \gamma_2(s)\nu^2 + \ldots\gamma_0(s). \tag{2.50}$$

The coefficients of the polynomials on the right-hand sides of these equations can be easily expressed in terms of known functions of the arc length s. It follows from Eq. (2.50) that the degree of the polynomial $\beta_1(s, \nu)$ is one less than the degree of $\alpha_3(s, \nu)$. The first equation then shows that $\alpha_3(s, \nu)$ is a third-degree polynomial

$$\alpha_3(s, \nu) = \alpha_{33}(s)\nu^3 + \alpha_{32}(s)\nu^2 + \alpha_{31}(s)\nu + \alpha_{30}(s). \tag{2.51}$$

Let us write $\beta_1(s, \nu)$ as

$$\beta_1(s, \nu) = \beta_{12}(s)\nu^2 + \beta_{11}(s)\nu + \beta_{10}(s). \tag{2.52}$$

Let us substitute the polynomials (2.51) and (2.52), as well as $\alpha_2(s, \nu)$ and $\beta_0(s, \nu)$ into Eq. (2.50) and then, equating the coefficients of like powers of ν to zero, we obtain

$$\begin{aligned}
\alpha_{33}(s) &= -\frac{1}{3}\frac{\beta_{12}'(s)}{c_0(s)\beta_{01}(s)} - \frac{4}{3}\alpha_{22}(s)\frac{\beta_{12}(s)}{\beta_{01}(s)} + \frac{1}{6i}\frac{1}{\beta_{01}(s)}\gamma_2(s),\\
\alpha_{32} &= -\frac{1}{2}\frac{\beta_{11}'(s)}{c_0(s)\beta_{01}} - \alpha_{22}(s)\frac{\beta_{11}(s)}{\beta_{01}(s)} - \alpha_{21}(s)\frac{\beta_{12}(s)}{\beta_{01}(s)} + \frac{1}{4i\beta_{01}(s)}\gamma_1(s),\\
\alpha_{31}(s) &= -\frac{\beta_{10}'}{c_0(s)\beta_{01}(s)} - \alpha_{21}\frac{\beta_{11}}{\beta_{01}} - \frac{\beta_{12}(s)}{i\beta_{01}(s)} + \frac{1}{2i\beta_{01}(s)}\gamma_0(s).
\end{aligned} \tag{2.53}$$

If we now substitute the known expressions for $\alpha_{3j}(s)$, j = 1, 2, 3, into Eq. (2.49), then by equating to zero the coefficients of ν^3, ν^2, and ν^1, we will obtain an equation for the determination of $\beta_{1l}(s)$, $l = 0, 1, 2$. By equating the free term to zero we can obtain $\alpha_{30}(s)$ in the form of an integral with a variable upper limit involving only known functions. If in the equations for $\beta_{1l}(s)$ we introduce new unknown functions $y_{1l}(s)$ defined by

$$\beta_{1l}(s) = \beta_{01}^{l+1}(s)\, c_0^{1/2}(s)\, y_{1l}(s) = \frac{y_{1l}(s)}{c_0^{l/2}(s)\, F^{l+1}(s)},$$

then these equations can be written as

$$y_{1l}''(s) + \left[K(s) + \frac{(l+1)^2 - 1}{4}\frac{1}{F^4(s)}\right] y_{1l}(s) = p_{1l}(s), \tag{2.54}$$

where

$$p_{12}(s) = \frac{3}{2}c_0^{1/2}F^2\left\{-2\frac{c_1c_2}{c_0^2} + \frac{1}{3}\frac{c_3}{c_0} + \frac{1}{\rho}\left[\frac{7}{3}\frac{c_2}{c_0} - \frac{2}{3}\left(\frac{1}{2}\frac{c_0'}{c_0} + \frac{F'}{F}\right)^2 + \frac{4}{3}\frac{\rho'}{\rho}\left(\frac{1}{2}\frac{c_0'}{c_0} + \frac{F'}{F}\right) - \frac{5}{6}\frac{1}{F^4}\right]\right\}.$$

The expression for $p_{11}(s)$ contains y(s), F(s), and $y_{12}(s)$ in addition to $c_0(s)$ and $\rho(s)$. After Eq. (2.54) with $l = 2$ has been integrated, the function $y_{12}(s)$ becomes known. It should be noted that $p_{11}(s) = 0$ when y(s) = 0. The function $p_{10}(s)$ depends on $c_0(s)$, $\rho(s)$, y(s), F(s), $y_{12}(s)$, and $y_{11}(s)$. If the first two equations of system (2.54) have been integrated, then $y_{12}(s)$ and $y_{11}(s)$ are known. The solution of the inhomogeneous Eq. (2.54) can be constructed by the Lagrange multiplier method. To do this, however, we first need to obtain the solution of the corresponding homogeneous equation*

$$f''(s) + \left[K(s) + \frac{(l+1)^2 - 1}{4}\frac{1}{F^4(s)}\right] f(s) = 0. \tag{2.55}$$

*Equation (2.55) for the special case K(s) = 0 was first integrated by Lazutkin [2].

It can be shown that the general solution of Eq. (2.55) can be easily obtained if the linearly independent solutions $y_1(s)$ and $y_2(s)$ of Eq. (2.44) and, consequently, the function F(s) are known. Let us introduce new variables in Eq. (2.55) defined by

$$t=\frac{1}{2}\int_0^s F^{-2}(\tau)\,d\tau, \quad \vartheta(s)=f(s)F^{-1}(s).$$

The function $\vartheta(s)$ satisfies the equation

$$\vartheta''(s)+(l+1)^2\vartheta(s)=0$$

and, consequently, we have

$$f(s)=F(s)\left\{A\sin\left(\frac{l+1}{2}\int_0^s\frac{d\tau}{F^2(\tau)}\right)+B\cos\left(\frac{l+1}{2}\int_0^s\frac{d\tau}{F^2(\tau)}\right)\right\}. \tag{2.56}$$

After the function $y_{1l}(s)$ have been found, the coefficients of the polynomial $\alpha_3(s,\nu)$ can be easily determined. The corresponding formulas for the case $\rho(s)=\infty$ and $y(s)=0$ are

$$\left.\begin{aligned}
\alpha_{33}(s)&=-\frac{1}{3}c_0^{-\frac{5}{2}}(s)F^{-2}(s)\left[\frac{y_{12}'(s)}{y_{12}(s)}-\frac{c_0'(s)}{c_0(s)}-\frac{F'(s)}{F(s)}\right]y_{12}(s),\\
\alpha_{32}(s)&=-\frac{1}{2}c_0^{-\frac{3}{2}}(s)F^{-1}(s)\left[\frac{y_{11}'(s)}{y_{11}(s)}-\frac{1}{2}\frac{c_0'(s)}{c_0(s)}-\frac{F'(s)}{F(s)}\right]y_{11}(s),\\
\alpha_{31}(s)&=-c_0^{-\frac{1}{2}}(s)\left[\frac{y_{10}'(s)}{y_{10}(s)}-\frac{F'(s)}{F(s)}\right]y_{10}(s)+ic_0^{-\frac{3}{2}}(s)F^{-2}(s)y_{12}(s),\\
\alpha_{30}&=\int_{C_3}^s\left[\frac{1}{2i}\left(\frac{y_{11}'(\tau)}{y_{11}(\tau)}-\frac{1}{2}\frac{c_0'(\tau)}{c_0(\tau)}-\frac{F'(\tau)}{F(\tau)}\right)-\frac{1}{F^2(\tau)}\right]\frac{y_{11}(\tau)}{c_0^{1/2}(\tau)F(\tau)}\,d\tau,
\end{aligned}\right\} \tag{2.57}$$

where C_3 is an arbitrary constant.

6. The polynomials $\alpha_n(s,\nu)$ and $\beta_{n-2}(s,\nu)$ for $n\geq 4$ can be found by a procedure analogous to that used above for the construction of the polynomials $\alpha_3(s,\nu)$ and $\beta_1(s,\nu)$. Assuming $m=n$ in system (2.21), we obtain the two equations

$$a_n(s,\nu)=0, \quad b_n(s,\nu)=0, \tag{2.58}$$

which, in addition to the unknown polynomials $\alpha_n(s,\nu)$ and $\beta_{n-2}(s,\nu)$, also contain the polynomials $\alpha_j(s,\nu)$, $j=0,1,\ldots,n-1$, and $\beta_{j-2}(s,\nu)$, $j=2,3,\ldots,n-1$, that have already been determined during the preceding stages. It can be shown that the polynomials $\alpha_n(s,\nu)$ and $\beta_{n-2}(s,\nu)$ are the n- and (n − 1)-degree polynomials

$$\alpha_n(s,\nu)=\alpha_{n,n}(s)\nu^n+\alpha_{n,n-1}(s)\nu^{n-1}+\ldots+\alpha_{n,0}(s),$$

$$\beta_{n-2}(s,\nu)=\beta_{n-2,n-1}(s)\nu^{n-1}+\beta_{n-2,n-2}(s)\nu^{n-2}+\ldots+\beta_{n-2,0}(s).$$

Substituting these polynomials into Eq. (2.58) and, as before, equating to zero the coefficients of like powers of ν, we obtain 2n + 1 equations for the determinations of the function $\alpha_{n,k}(s)$, $k=0,1,\ldots,n$, and $\beta_{n-2,l}(s)$, $l=0,1,\ldots,n-1$. If we introduce new unknown functions

$y_{n-2,l}(s)$ instead of $\beta_{n-2,l}(s)$ with the help of the formulas

$$\beta_{n-2,l}(s)=c_0^{-\frac{l}{2}}(s)\,F^{-(l+1)}(s)\,y_{n-2,l}(s),$$

then $y_{n-2,l}(s)$ satisfies an equation analogous to (2.54)

$$y''_{n-2,l}(s)+\left[K(s)+\frac{(l+1)^2-1}{4}\frac{1}{F^4(s)}\right]y_{n-2,l}(s)=p_{n-2,l}, \qquad (2.59)$$

$$l=0,\ 1,\ \ldots,\ n-1,$$

where $p_{n-2,n-1}(s)$ is a known function and $p_{n-2,l}(s)$ for $l < n-1$ can be obtained from the $y_{n-2,j}(s)$, $j < l$, found during the solution of the preceding equations. The general solution of Eq. (2.59) can be constructed in the analogous manner to the general solution of (2.54) by the Lagrange multiplier method applied to the general solution (2.56) of the homogeneous equation (2.55). After the $y_{n-2,l}(s)$ have been found, the $\alpha_{n,k}(s)$ can be determined from formulas analogous to (2.57), the $\alpha_{n,0}(s)$ in the same way as $\alpha_{3,0}(s)$ in (2.57) being given by a quadrature containing an arbitrary constant.

Having calculated $\alpha_n(s,\nu)$ and $\beta_{n-2}(s,\nu)$ by the above method, we can construct a function of the form of (2.12) whose substitution into the Helmholtz equation (2.5) leads to a residual (difference between the left-hand side and zero) of order $\omega^{-(n-3)/2}$. By a suitable choice of n, we can make the residual as small as we like. The constants arising from the integration of the differential equations are determined from the boundary conditions which must be formulated in some manner when the curve of the problem is specified.

§ 3. The Asymptotic Behavior of the Eigenfunctions and Eigenvalues of the Boundary Problem for the Waveguide

The present section is devoted to the investigation of the boundary problem concerning the natural oscillations of a bounded waveguide in an inhomogeneous medium. In the construction of the eigenfunctions of the waveguide and the determination of the corresponding eigenvalues, we will make use of the solutions of the wave equation concentrated near the waveguide axis obtained in the preceding section.

1. Let a curve l be the axis of a waveguide which is stable in the first approximation and which is bounded by two perfectly reflecting walls (see Fig. 1). As in the preceding section, we will introduce curvilinear coordinates (s, n) in the vicinity of the curve l. We will measure the arc length s along the curve from the left end of the waveguide and we will take the length of the waveguide to be equal to d. Let us look for solutions of the wave equation

$$\Delta u+\frac{\omega^2}{c^2(s,n)}u=0 \qquad (3.1)$$

concentrated near the waveguide axis, i.e., solutions satisfying the conditions

$$u(s,n)\to 0 \quad \text{as} \quad |n\omega^{1/2}|\to\infty \qquad (3.2)$$

and becoming zero when s = 0 and s = d, i.e.,

$$u(0,n)=u(d,n)=0. \qquad (3.3)$$

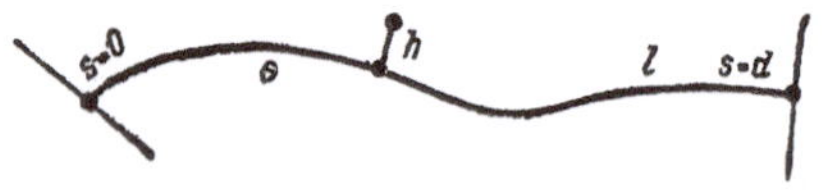

Fig. 1

In addition to finding the functions u(s, n), we will also have to determine the eigenvalues of the problem (3.1)-(3.3). In the preceding section we have constructed solutions of the wave equation (3.1) concentrated in the vicinity of l in the form

$$u(s,\ n)=\exp\left\{i\omega\sum_{n=0}^{\infty}\alpha_n(s,\ \nu)\,\omega^{-\frac{n}{2}}\right\}D_q\left\{\sum_{n=0}^{\infty}\beta_n(s,\ \nu)\,\omega^{-\frac{n}{2}}\right\}.$$

The polynomials $\alpha_n(s, \nu)$ and $\beta_n(s, \nu)$ were determined apart from arbitrary constants with the help of a recurrence system consisting of differential equations. We will show that the boundary conditions (3.2) allow us to determine these constants and, moreover, they lead to an equation which can be used to find the eigenvalues of the problem.

We will seek the eigenfunctions U(s, ν) of problem (3.1)-(3.3) in the form

$$U(s,\ n)=Au(s,\ n)+Bu^*(s,\ n), \tag{3.4}$$

where A and B are two constants; the asterisk denotes a complex conjugate. Substituting (3.4) into the boundary condition (3.3), we obtain a system of two equations for the determination of A and B

$$\begin{aligned} &Au(0,\ n)+Bu^*(0,\ n)=0,\\ &Au(d,\ n)+Bu^*(d,\ n)=0, \end{aligned} \tag{3.5}$$

which after the cancellation of the factor

$$\exp\left\{-\omega\,\mathrm{Im}\sum_{n=0}^{\infty}\alpha_n(s,\ \nu)\,\omega^{-\frac{n}{2}}\right\}$$

can be written as

$$\begin{aligned} &A\exp\left\{i\omega\,\mathrm{Re}\sum_{n=0}^{\infty}\alpha_n(s,\ \nu)\,\omega^{-\frac{n}{2}}\right\}D_q\left\{\sum_{n=0}^{\infty}\beta_n(s,\ \nu)\,\omega^{-\frac{n}{2}}\right\}+\\ &+B\exp\left\{-i\omega\,\mathrm{Re}\sum_{n=0}^{\infty}\alpha_n^*(s,\ \nu)\,\omega^{-\frac{n}{2}}\right\}D_q\left\{\sum_{n=0}^{\infty}\beta_n^*(s,\ \nu)\,\omega^{-\frac{n}{2}}\right\}\Bigg|_{s=0,\ s=d}=0. \end{aligned} \tag{3.6}$$

The constants A and B are independent of ν if the coefficients of the system after the cancellation factors are independent of ν. Let us require that the following relations hold when s = 0 and s = d:

$$\beta_n(s,\ \nu)=\beta_n^*(s,\ \nu),\quad n=0,\ 1,\ 2\ldots,$$

i.e.,

$$\mathrm{Im}\,\beta_n(0,\ \nu)=\mathrm{Im}\,\beta_n(d,\ \nu)=0,\quad n=0,\ 1,\ 2\ldots, \tag{3.7}$$

which allow us to remove the parabolic cylinder function from the equations of system (3.7). Since the $\beta_n(s, \nu)$ are polynomials in

$$\beta_n(s, \nu) = \beta_{n,\, n+1}(s)\,\nu^{n+1} + \beta_{n,\, n}(s)\,\nu^n + \ldots + \beta_{n,\, 0}(s),$$

it is obvious that equality (3.7) is equivalent to the conditions

$$\begin{gathered} \operatorname{Im} \beta_{n,\, k}(0) = \operatorname{Im} \beta_{n,\, k}(d) = 0, \quad n = 0,\ 1,\ 2\ \ldots, \\ k = 0,\ 1,\ \ldots,\ n+1. \end{gathered} \tag{3.8}$$

The arguments of the exponential functions are independent of ν if we take the real parts of the polynomials

$$\alpha_n(s, \nu) = \alpha_{n,\, n}(s)\,\nu^n + \alpha_{n,\, n-1}(s)\,\nu^{n-1} + \ldots + \alpha_{n,\, 0}(s)$$

to be independent of ν at $s = 0$ and $s = d$, i.e., if we take

$$\begin{gathered} \operatorname{Re} \alpha_{n,\, k}(0) = \operatorname{Re} \alpha_{n,\, k}(d) = 0, \quad n = 1,\ 2,\ \ldots, \\ k = 1,\ 2,\ \ldots,\ n. \end{gathered} \tag{3.9}$$

When (3.8) and (3.9) are satisfied, the system of equations (3.6) reduces to

$$A \exp\left\{ i\omega \operatorname{Re} \sum_{n=0}^{\infty} \alpha_{n,\, 0}(s)\, \omega^{-\frac{n}{2}} \right\} + B \exp\left\{ -i\omega \operatorname{Re} \sum_{n=0}^{\infty} \alpha_{n,\, 0}(s)\, \omega^{-\frac{n}{2}} \right\} \Bigg|_{s=0,\, d} = 0. \tag{3.10}$$

It should be recalled that the free terms of the polynomials $\alpha_n(s, \nu)$, i.e., the function $\alpha_{n,0}(s)$, have been expressed as integrals with a variable upper limit s and an arbitrary lower limit. By a suitable choice of A and B we can always secure that the equalities

$$\alpha_{n,\, 0}(0) = 0, \quad n = 0,\ 1,\ 2,\ \ldots. \tag{3.11}$$

are satisfied. In the following, we will assume that (3.11) always holds. When (3.11) are taken into account, the system (3.10) assumes the simple form

$$\begin{gathered} A + B = 0, \\ A \exp\left\{ i\omega \operatorname{Re} \sum_{n=0}^{\infty} \alpha_{n,\, 0}(d)\, \omega^{-\frac{n}{2}} \right\} + B \exp\left\{ -i\omega \operatorname{Re} \sum_{n=0}^{\infty} \alpha_{n,\, 0}(d)\, \omega^{-\frac{n}{2}} \right\} = 0. \end{gathered}$$

In order that this homogeneous system has a nontrivial solution, it is necessary for its determinant

$$\Delta = -2i \sin\left\{ \omega \operatorname{Re} \sum_{n=0}^{\infty} \alpha_{n,\, 0}(d)\, \omega^{-\frac{n}{2}} \right\}$$

to be different from zero. The equation $\Delta = 0$ represents the equation for the determination of the eigenvalues of the problem. It is obvious that the equation $\Delta = 0$ is equivalent to

$$\omega \operatorname{Re} \sum_{n=0}^{\infty} \alpha_{n,\, 0}(d)\, \omega^{-\frac{n}{2}} = \pi p, \tag{3.12}$$

where p is an integer. We must assume that $p \gg 1$, because all derivations refer to the case of sufficiently large ω.

Let us return to conditions (3.9). When $n = 1$, it is satisfied automatically since $\alpha_1(s, \nu) = C_1$ [see formula (2.32)]. In view of (3.11), the constant C_1 should be taken equal to zero. Hence, we see that $\alpha_1(s, \nu) \equiv 0$. Assuming that $n = 2$ in (3.9), we obtain $\operatorname{Re} \alpha_{2,1}(s)|_{s=0,\, s=d} = 0$ and $\operatorname{Re} \alpha_{2,2}(s)|_{s=0,\, s=d} = 0$. Since we have $\alpha_{21} = -(1/c_0)(\beta_{00}'/\beta_{01})$, $\alpha_{22} = -(\frac{1}{2} c_0)(\beta_{01}'/\beta_{01})$, and conditions (3.8) must hold, we find that

$$\operatorname{Re} \beta_{0,0}'(0) = \operatorname{Re} \beta_{0,1}'(d) = 0, \quad \operatorname{Re} \beta_{0,1}'(0) = \operatorname{Re} \beta_{0,1}'(d) = 0. \tag{3.13}$$

We have seen that $F(s) = c_0^{-1/2}(s) \cdot \beta_{01}^{-1}(s)$ must satisfy the nonlinear Eq. (2.43), namely,

$$F'' + K(s) F = \frac{1}{4} \frac{1}{F^3}. \tag{3.14}$$

It follows from (3.8) and (3.13) that this equation must be solved with the following boundary conditions:

$$\operatorname{Re}\left(\frac{F'}{F} + \frac{1}{2} \frac{c_0'}{c_0}\right)\Bigg|_{s=0,\, s=d} = 0 \text{ and } \operatorname{Im} F|_{s=0,\, s=d} = 0.$$

Since $K(s)$ is real, we can assume that $F(s)$ is also real and that it satisfies Eq. (3.14) and the boundary conditions

$$\frac{F'}{F} + \frac{1}{2} \frac{c_0'}{c_0}\Bigg|_{s=0,\, s=d} = 0. \tag{3.15}$$

We have also seen that $y(s) = \beta_{00}(s)F(s)$ must satisfy the homogeneous second-order equation

$$y'' + K(s) y = 0. \tag{3.16}$$

Conditions (3.8), (3.13) and Eq. (3.16) will only be satisfied if $y(s) = 0$ [we are not here interested in the case when Eq. (3.16) has a nonzero solution satisfying the homogeneous conditions (3.8) and (3.13), i.e., when $\beta_{0,0}(s) \neq 0$]. Let us now consider the case $n = 3$. It has been shown in the preceding section that the functions $\alpha_{3j}(s)$, $j = 1, 2, 3$, can be expressed in terms of the functions $\beta_{1l}(s)$, $l = 0, 1, 2$ [formulas (2.53)]. Thus, the conditions (3.9) are in effect imposed on the $\beta_{1l}(s)$ and with (3.8) taken into account, they reduce to

$$\operatorname{Re} \beta_{1l}'(0) = \operatorname{Re} \beta_{1l}'(d) = 0, \quad l = 0, 1, 2. \tag{3.17}$$

It can be shown that in the general case with $n \geq 4$, conditions (3.9) reduce to

$$\operatorname{Re} \beta_{n-2,\, l}'(s)|_{s=0,\, s=d} = \delta_{n-2,\, l}^{(0,\, d)}, \quad \begin{matrix} n = 4, 5 \ldots \\ l = 0, 1 \ldots n-1, \end{matrix} \tag{3.18}$$

where $\delta_{n-2,\, l}^{(0,\, d)}$ are constants that are found from the values at $s = 0$ and $s = d$ of the polynomials $\beta_k(s, \nu)$, $k = 0, 1, \ldots, n-1$, and $\alpha_k(s, \nu)$, $k = 0, 1, \ldots, n-1$, derived during the preceding stages.

In deriving $\beta_{n-2,\, l}(s)$, we have introduced new unknown functions $y_{n-2,\, l}(s) = c_0^{l/2}(s) F^{l+1} \times (s) \beta_{n-2,\, l}(s)$ and have obtained Eqs. (2.54) satisfied by them. It follows from (3.17) and (3.18) that at the end-points of the interval [0, d], the function $y_{n-2,\,l}(s)$ must satisfy the conditions

$$\operatorname{Re}\left\{y'_{n-2,\,l}(s)-\left[\frac{l}{2}\frac{c_0'(s)}{c_0(s)}+(l+1)\frac{F'(s)}{F(s)}\right]y_{n-2,\,l}(s)\right\}\Big|_{s=0,\ s=d}=$$

$$=\delta_{n-2,\,l}^{(0,\,d)}c_0^{l/2}(s)\,F^{l+1}(s)\big|_{s=0,\ s=d},\qquad \begin{matrix} n=3,\ 4,\ 5,\ \ldots \\ l=0,\ 1,\ \ldots\ n-1. \end{matrix} \tag{3.19}$$

Since F(s) in turn must satisfy conditions (3.15), expressions (3.19) can be rewritten as

$$\operatorname{Re}\left\{y'_{n-2,\,l}(s)+\frac{1}{2}\frac{c_0'(s)}{c_0(s)}y_{n-2,\,l}(s)\right\}\Big|_{s=0,\ s=d}=\delta_{n-2,\,l}^{(0,\,d)}c_0^{l/2}F^{l+1}(s)\big|_{s=0,\ s=d}. \tag{3.20}$$

These equalities must be supplemented by the conditions for the imaginary parts

$$\operatorname{Im} y_{n-2,\,l}(s)\big|_{s=0,\ s=d}=0, \tag{3.21}$$

which follow from the requirements specified by (3.8).

Thus, Eqs. (2.54) must be solved with the boundary conditions (3.20) and (3.21). Equations (2.54) and conditions (3.20) may be inhomogeneous as well as homogeneous. Let us establish the conditions under which the homogeneous equation (2.54) has a nontrivial solution satisfying the homogeneous boundary conditions (3.20) and (3.21). As we have seen, the solution of the homogeneous equation (2.54) can be written as

$$y_{n-2,\,l}(s)=F(s)\left\{A\sin\left(\frac{l+1}{2}\int_0^s F^{-2}(\tau)\,d\tau\right)+B\cos\left(\frac{l+1}{2}\int_0^s F^{-2}(\tau)\,d\tau\right)\right\}.$$

Substituting $y_{n-2,\,l}(s)$ into the homogeneous conditions (3.20) and (3.21) and taking the boundary conditions (3.15) into account, we obtain a system of equations for the constants A and B

$$\operatorname{Re}\frac{l+1}{2}\frac{1}{F(s)}\left\{A\cos\left(\frac{l+1}{2}\int_0^s F^{-2}(\tau)\,d\tau\right)-B\sin\left(\frac{l+1}{2}\int_0^s F^{-2}(\tau)\,d\tau\right)\right\}\Big|_{s=0,\ s=d}=0,$$

$$\operatorname{Im}\left\{A\sin\left(\frac{l+1}{2}\int_0^s F^{-2}(\tau)\,d\tau\right)+B\cos\left(\frac{l+1}{2}\int_0^s F^{-2}(\tau)\,d\tau\right)\right\}\Big|_{s=0,\ s=d}=0.$$

Setting s = 0, we find that Re A = Im B = 0. For s = d, we then obtain

$$\operatorname{Re} B\sin\left(\frac{l+1}{2}\int_0^d F^{-2}(\tau)\,d\tau\right)=0,$$

$$\operatorname{Im} A\sin\left(\frac{l+1}{2}\int_0^d F^{-2}(\tau)\,d\tau\right)=0.$$

If

$$\frac{1}{2}\int_0^d F^{-2}(\tau)\,d\tau\neq\frac{m}{l+1}\pi, \tag{3.22}$$

where m is an integer, then we have Re B = Im A = 0. When

$$\frac{1}{2}\int_0^d F^{-2}(\tau)\,d\tau=\frac{m}{l+1}\pi\,, \tag{3.23}$$

Re B and Im A remain arbitrary. It is obvious that when conditions (3.22) holds, the corresponding inhomogeneous problem (2.54) (3.20), (3.21) is soluble, whereas when (3.23) holds it does not have a solution. It follows from this that the process of construction of the polynomials $\alpha_n(s, \nu)$ and $\beta_n(s, \nu)$ may be continued indefinitely when

$$\frac{1}{2}\int_0^d F^{-2}(\tau)\,d\tau\neq\frac{p}{q}\pi, \tag{3.24}$$

where p/q is an irrational fraction. In the converse case, an insoluble problem arises in the determination of the leading coefficient $\beta_{q-2,\,q-1}(s)$ of the polynomial $\beta_{q-2}(s)$ and, in general, this polynomial cannot be constructed by means of the procedure outlined above. Lazutkin [2] was the first to point out the possibility of a situation of this type which he encountered in the construction of "bouncing-ball" eigenfunctions in a plane.

2. The preceding section contains a procedure by which it is possible to construct all the polynomials $\alpha_n(s, \nu)$ and $\beta_n(s, \nu)$, n = 0, 1, 2, ... If the polynomials $\alpha_n(s, \nu)$ and $\beta_n(s, \nu)$ are known, then Eqs. (3.12) can be used to determine the natural frequencies $\omega_{q,\,p}$ to as high an accuracy as we wish and we can then construct the eigenfunction $U_{q,p}(s, n)$ satisfying Eq. (3.1) and the boundary conditions (3.3) to as high an accuracy as we wish.

In the present section, we will restrict ourselves to the construction of the first approximation for the eigenfunctions and eigenfrequencies. It follows from Eq. (3.12) that*

$$\omega_{q,\,p}=\frac{1}{\operatorname{Re}\alpha_{00}(d)}\left[\pi p-\operatorname{Re}\alpha_{20}(d)+O\left(\frac{1}{p^{1/2}}\right)\right]. \tag{3.25}$$

Substituting the values of $\alpha_{00}(d)$ and $\alpha_{20}(d)$ into formula (3.25), we obtain

$$\omega_{q,\,p}=\left(\int_0^d\frac{d\tau}{c_0(\tau)}\right)^{-1}\left[\pi p+\frac{q+1/2}{2}\int_0^d F^{-2}(\tau)\,d\tau+O\left(\frac{1}{p}\right)\right]. \tag{3.26}$$

Let us evaluate $\int_0^d F^{-2}(\tau)\,d\tau$. Let us make use of formula (2.46)

$$\frac{1}{2}\int_0^d\frac{d\tau}{F^2(\tau)}=\frac{1}{2}\int_0^d\frac{d\tau}{a_{11}y_1^2(\tau)+2a_{12}y_1(\tau)\,y_2(\tau)+a_{22}y_2^2(\tau)}.$$

Let us introduce a new integration variable $x = y_2(\tau)/y_1(\tau)$. Then, we have $d\tau = y_1^2(\tau)/W[y_1, y_2]$ and, in view of the fact that the Wronskian $W[y_1, y_2]$ is a constant, we obtain

$$\frac{1}{2}\int_0^d\frac{d\tau}{F^2(\tau)}=\frac{1}{2\Delta W}\arctan\frac{a_{22}y_2(\tau)\,y_1^{-1}(\tau)+a_{12}}{\Delta}\Bigg|_0^d+\pi N, \tag{3.27}$$

* It can be shown that the actual error in formula (3.25) is of order 1/p.

where $\Delta^2 = a_{11}a_{22} - a_{12}^2$ and N is the number of zeros of $y_1(s)$ inside the interval $[0, d]$. The elements of the symmetric matrix $\| a_{k,j} \|$ must be chosen such that condition (3.15) is satisfied. Substituting (2.46) into (3.15) we arrive at the two equations

$$\sum_{k,\, j=1}^{2} [y_j'(s) + \varkappa(s)\, y_j(s)]\, a_{kj} y_k(s) |_{s=0,\ s=d} = 0, \tag{3.28}$$

where

$$\varkappa(s) = c_0'(s)\, c_0^{-1}(s).$$

We choose the linearly independent solutions of (2.44), namely, $y_1(s)$ and $y_2(s)$, such that the following equalities are satisfied:

$$\begin{aligned} y_1'(0) + \varkappa(0)\, y_1(0) &= 0, \\ y_2'(d) + \varkappa(d)\, y_2(d) &= 0. \end{aligned} \tag{3.29}$$

In view of (3.29), Eqs. (3.28) can be rewritten as

$$\begin{aligned} a_{12} y_1(0) + a_{22} y_2(0) &= 0, \\ a_{11} y_1(d) + a_{21} y_2(d) &= 0. \end{aligned} \tag{3.30}$$

Making use of (3.30), as well as equality (2.45) which implies that $\Delta^2 W^2 = 1/4$, we can transform formula (3.27) into

$$\frac{1}{2}\int_0^d \frac{d\tau}{F^2(\tau)} = \arctan\left[\operatorname{sign}\left(W \frac{y_1(0)}{y_2(0)}\right) \sqrt{\frac{y_1(0)\, y_2(d) - y_1(d)\, y_2(0)}{y_1(d)\, y_2(0)}}\right] + \pi N,$$

where

$$\frac{1}{2}\int_0^d \frac{d\tau}{F^2(\tau)} = \operatorname{sign}\left(\varkappa(0) + \frac{y_2'(0)}{y_2(0)}\right) \arccos \sqrt{\frac{y_1(d)}{y_1(0)}\, \frac{y_2(0)}{y_2(d)}} + \pi N. \tag{3.31}$$

Finally, we find the following expression for the eigenfrequencies $\omega_{q,p}$ when $p \gg 1$:

$$\omega_{q,p} = \left(\int_0^d \frac{d\tau}{c_0(\tau)}\right)^{-1} \left[\pi p + \left(q + \frac{1}{2}\right)\left(\operatorname{sign}\left(\varkappa(0) + \frac{y_2'(0)}{y_2(0)}\right) \arccos \sqrt{\frac{y_1(d)\, y_2(0)}{y_1(0)\, y_2(d)}} + \pi N\right) + O\left(\frac{1}{p}\right)\right]. \tag{3.32}$$

It should be recalled that $y_1(s)$ and $y_2(s)$ have been chosen such that conditions (3.29) are satisfied.

Let us substitute the eigenvalues found above into formula (3.4). Taking into account that $A = -B$, the expression for the eigenfunctions $U_{q,p}$ in the first approximation is

$$U_{q,p}(s, n) \sim \sqrt{\frac{c_0^{1/2}(s)}{F(s)}} \left\{ \sin\left(\omega_{q,p} \cdot \int_0^s c_0^{-1}(\tau)\, d\tau - \frac{q + \frac{1}{2}}{2} \int_0^s F^{-2}(\tau)\, d\tau + \right.\right.$$

$$\left.\left. + \left[\frac{1}{2}\frac{c_0'(s)}{c_0(s)} + \frac{F'(s)}{F(s)}\right] \frac{\nu^2}{2c_0(s)}\right) + O\left(\frac{1}{\omega_{q,p}^{1/2}}\right)\right\} D_q\left\{\frac{\nu}{c_0^{1/2}(s)\, F(s)}\right\} + D_q'\left\{\frac{\nu}{c_0^{1/2}(s)\, F(s)}\right\} O\left(\frac{1}{\omega_{q,p}^{1/2}}\right).$$

3. The above procedure for the construction of the asymptotic expressions for the eigenfunctions concentrated near the axis of a bounded waveguide can also be used for the determination of the eigenfunctions concentrated in the vicinity of stable diameters of two-dimensional regions. Let us conformally map the two-dimensional region onto a band. Under the mapping, the stable diameter transforms into the waveguide axis and we obtain the problem concerning the eigenfunctions concentrated in the vicinity of the axis of a bounded waveguide. With this approach, the problem of the eigenfunctions concentrated in the vicinity of the diameter of the region formed by two circular arcs (see Fig. 2) becomes particularly simple.

Performing a conformal mapping of the lune shown in Fig. 2 onto a band of width π, we obtain the problem of the eigenfunctions of a waveguide in an inhomogeneous medium in which the wave-propagation velocity is

$$c(s,\ n)=\frac{\pi}{a\beta}\left[\cosh\frac{\beta}{\pi}n-\cos\left(\frac{\beta}{\pi}s+\alpha\right)\right],\quad 0\leqslant s\leqslant\pi.$$

With this expression for the velocity and $\alpha > \beta$, we can rewrite formula (3.23) for the eigenfrequencies as

$$\omega_{q,\,p}=\left(\frac{a\beta}{2\pi}\int\limits_0^{\pi}\frac{d\tau}{\sin^2\frac{1}{2}\left(\frac{\beta}{\pi}\tau+\alpha\right)}\right)^{-1}\left[\pi p+\left(q+\frac{1}{2}\right)\arccos\sqrt{\frac{\sin\frac{\alpha-\beta}{2}\ \sin\frac{\alpha+2\beta}{2}}{\sin\frac{\alpha}{2}\ \sin\frac{\alpha+\beta}{2}}}+O\left(\frac{1}{p}\right)\right].$$

Taking the relations between the angles α and β and between the radii R_1 and R_2 of the circular arcs into account, we obtain

$$\omega_{q,\,p}=\frac{1}{2d}\left[\pi p+\left(q+\frac{1}{2}\right)\arccos\sqrt{\left(1-\frac{2d}{R_1}\right)\left(1-\frac{2d}{R_2}\right)}+O\left(\frac{1}{p}\right)\right], \tag{3.33}$$

where 2d is the diameter of the lune. This formula for the eigenfrequencies is well known in the theory of open resonators. We can obtain the subsequent terms with the help of the polynomials $\alpha_3(s,\ \nu)$, $\alpha_4(s,\ \nu)$, etc.

In conclusion, the author would like to express his gratitude to all participants of the seminar on wave diffraction at the Leningrad Branch of the Steklov Mathematical Institute and

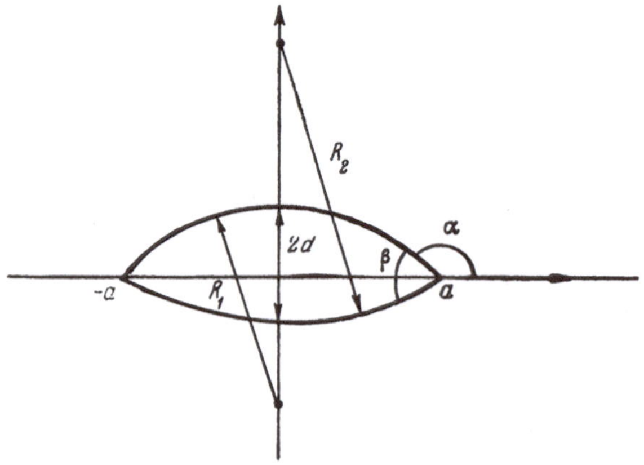

Fig. 2

the Leningrad State University and particularly to V. M. Babich for valuable discussions and advice.

LITERATURE CITED

1. V. M. Babich and V. F. Lazutkin, "Eigenfunctions concentrated near a closed geodesic," in: Topics in Mathematical Physics, Vol. 2, Consultants Bureau, New York (1968).
2. V. F. Lazutkin, "Construction of an asymptotic series for eigenfunctions of the 'bouncing-ball' type," Trudy MIAN, Vol 95 (in press).
3. E. Kamke, Handbook of Ordinary Differential Equations [Russian translation], IL (1950), p. 199.
4. V. A. Yakubovich, "The stability of the solutions of a system of two canonical linear equations with periodic coefficients," Matem. Sbornik, Vol. 37 (79), 1 (1955).

PERTURBATIONS OF THE SPECTRUM OF THE SCHROEDINGER OPERATOR WITH A COMPLEX PERIODIC POTENTIAL

V. A. Zheludev

The present article is devoted to an investigation of the discrete spectrum of the nonself-adjoint Schroedinger operator with a complex periodic potential perturbed by a decreasing potential. It is well known [1] that the spectrum of the unperturbed operator is a purely continuous spectrum situated on a denumerable sequence of arcs in the λ plane. This sequence is contained in a half-strip $|\operatorname{Im}\lambda| < M_1$, $\operatorname{Re}\lambda > M_2$ which extends to infinity. The principal result of this article is contained in Section 2 and can be stated as follows: Under the condition

$$q(x) < C\exp(-\delta|x|^{1/2})$$

[$q(x)$ is the perturbing potential and $\delta > 0$], the set of the eigenvalues of the perturbed potential does not have accumulation points at a finite distance from the origin.* In proving this result, we make use of results obtained by Pavlov [2] in his investigation of the analogous problem in the absence of a periodic potential. The required result can be obtained in a much simpler way if we assume a condition of the form

$$q(x)\exp(\delta|x|)\in L(-\infty,\ \infty)\ (\delta > 0).$$

The corresponding proof is given in Section 3. It is shown in Section 4 that the continuous spectrum of the perturbed operator does not carry any eigenvalues.

§ 1. Preliminary Information

In the present section we will present the necessary facts concerning the unperturbed operator, obtain estimates for the derivatives of the kernel of the operator resolvent, as well as prove an auxiliary lemma concerning analytic functions.

Let $p_1(x)$ and $p_2(x)$ be real piecewise continuous functions with unit period, let $p(x) = p_1(x) + ip_2(x)$, and let L_0 be the Schroedinger operator in $L_2(-\infty, \infty)$ defined by $L_0u = -u'' + p(x)u$.

---

* It can be shown that the eigenvalues cannot accumulate toward the end points of the continuous spectrum under the weaker condition

$$\int_{-\infty}^{\infty} |q(x)|(1+x^2)\,dx < \infty.$$

Let us consider the solutions $\varphi(x, \lambda)$ and $\theta(x, \lambda)$ of the equation

$$L_0 u = \lambda u \tag{1.1}$$

satisfying the boundary conditions

$$\varphi(0, \lambda) = 0, \quad \theta(0, \lambda) = 1,$$
$$\varphi'_x(0, \lambda) = 1, \quad \theta'_x(0, \lambda) = 0.$$

Let us introduce the following abbreviations:

$$\begin{aligned} &\varphi(1, \lambda) \equiv \varphi(\lambda), \quad \theta(1, \lambda) \equiv \theta(\lambda), \\ &\varphi'_x(1, \lambda) \equiv \varphi'(\lambda), \quad \theta'_x(1, \lambda) \equiv \theta'(\lambda), \\ &\frac{1}{2}[\varphi'(\lambda) + \theta(\lambda)] \equiv F(\lambda). \end{aligned} \tag{1.2}$$

The functions $\varphi(x, \lambda)$ and $\theta(x, \lambda)$ are linearly independent and their Wronskian is equal to unity. It follows from this that

$$\theta(\lambda)\varphi'(\lambda) - \theta'(\lambda)\varphi(\lambda) \equiv 1. \tag{1.3}$$

The functions $\varphi(x, \lambda)$ and $\theta(x, \lambda)$ and, consequently, the functions defined by formulas (1.2) are integral functions of λ.

Let us consider two solutions of Eq. (1.1), say,

$$\begin{aligned} \psi_1(x, \lambda) &\equiv \theta(x, \lambda) + m_1 \varphi(x, \lambda), \\ \psi_2(x, \lambda) &\equiv \theta(x, \lambda) + m_2 \varphi(x, \lambda), \end{aligned} \tag{1.4}$$

and let us prove the following assertion.

LEMMA 1. If in formula (1.4) we have

$$\begin{aligned} m_1(\lambda) &= \frac{\varphi'(\lambda) - \theta(\lambda)}{2\varphi(\lambda)} + \frac{\sqrt{F^2(\lambda) - 1}}{\varphi(\lambda)}, \\ m_2(\lambda) &= \frac{\varphi'(\lambda) - \theta(\lambda)}{2\varphi(\lambda)} - \frac{\sqrt{F^2(\lambda) - 1}}{\varphi(\lambda)}, \end{aligned} \tag{1.5}$$

then the functions $\varphi(x, \lambda)$ have the following representations

$$\begin{aligned} \psi_1(x, \lambda) &= \rho^x \chi_1(x, \lambda), \\ \psi_2(x, \lambda) &= \rho^{-x} \chi_2(x, \lambda), \end{aligned} \tag{1.6}$$

where

$$\rho = F(\lambda) + \sqrt{F^2(\lambda) - 1}. \tag{1.7}$$

and $x_i(x, \lambda)$ are periodic functions with unit period. (Here, and everywhere below, we have in mind the branch of $\sqrt{z^2 - 1}$ which is positive for $z > 1$.)

PROOF. Let us first of all note that

$$m_1(\lambda) m_2(\lambda) = -\frac{\theta'(\lambda)}{\varphi(\lambda)}, \tag{1.8}$$

which follows directly from formulas (1.5) and (1.3). Let us now consider the function $x_1 \equiv \psi_1 \rho^{-x}$ and let us show that it is periodic in x with unit period. It can be seen from the definitions of $\psi_1(x, \lambda)$ and $m_1(\lambda)$ that

$$\chi_1(0, \lambda) = \psi_1(0, \lambda) = \theta(0, \lambda) = 1,$$
$$\chi'_{1x}(0, \lambda) = \psi'_{1x}(0, \lambda) - \psi_1(0, \lambda) \ln \rho = m_1 \varphi'_x(0, 1) - \ln \rho = m_1 - \ln \rho.$$

At the same time, the calculation of $\chi_1(x, \lambda)$ and its derivatives at $x = 1$ yields

$$\chi_1(1, \lambda) = \psi_1(1, \lambda)\rho^{-1} = [\theta(\lambda) + m_1(\lambda)\varphi(\lambda)]\rho^{-1} = 1,$$
$$\chi'_{1x}(1, \lambda) = [\psi'_{1x}(1, \lambda) - \psi_1(1, \lambda)\ln\rho]\rho^{-1} =$$
$$= [\theta'(\lambda) + m_1\varphi'(\lambda)]\rho^{-1} - \ln\rho = [-m_1 m_2\varphi(\lambda) + m_1\varphi'(\lambda)]\rho^{-1} - \ln\rho =$$
$$= m_1[m_2\varphi(\lambda) + \varphi'(\lambda)]\rho^{-1} - \ln\rho = m_1 - \ln\rho.$$

Thus, we have

$$\chi_1(0, \lambda) = \chi_1(1, \lambda), \quad \chi'_1(0, \lambda) = \chi'_{1x}(1, \lambda).$$

Since by definition we have $p(x + 1) = p(x)$, we see that $\psi_1(x, \lambda)$ is a periodic function of x with unit period. In a similar manner it can be shown that $\chi_2 = \psi_2\rho^x$ is a periodic function.

Let us now make a few remarks concerning the nature of the spectrum of the operator L_0. Let us consider in the interval [0, 1] the operator $\tilde{L}_t$ generated by the differential operation $lu = -u'' + p(x)u$ and the boundary conditions

$$u(1) = e^{it}u(0), \quad u'(1) = e^{it}u'(0), \quad t \in [0, 2\pi). \tag{1.9}$$

$\tilde{L}_t$ is a regular operator and possesses a denumerable sequence of eigenvalues $\lambda_n(t)$ extending to infinity. It is well known that the spectral arc of L_0 of number n is described by the equation $\lambda = \lambda_n(t)$ with $t \in | 0, 2\pi)$. It can be seen from this that the spectrum of L_0 consists of those λ for which $|\rho(\lambda)| = 1$. We can write the equation of the arc as follows: $\rho(\lambda) = e^{it}$. We therefore see that the function $\psi_1(x, \lambda_n)$ is an eigenfunction of $\tilde{L}_t$ for a t which satisfies $\lambda_n(t) = \lambda_n$. However, for the same λ_n there exists a function $\psi_2(x, \lambda_n)$ such that it is clearly the eigenfunction of the operator $\tilde{L}_{-t}$ or, what is the same, of the operator $\tilde{L}_{2\pi-t}$. Therefore, we see that $\lambda_n(t) = \lambda_n(2\pi - t)$ and, consequently, the end points of the arc are the λ'_n and λ''_n such that $\lambda'_n = \lambda_n(0)$ and $\lambda''_n = \lambda_n(\pi)$, i.e., the eigenvalues of the periodic and anti-periodic problems.

Let us now select any one arc of the spectrum and in this and the following sections we will only consider the neighborhood of this arc. It is clear from the above that $\rho(\lambda)$ is a function which realizes a one-to-one mapping of the spectral arc onto the unit circle. This mapping is carried out with the help of the function $F(\lambda)$ which is related to ρ as follows:

$$\rho(F) = F + \sqrt{F^2 - 1},$$
$$F(\rho) = \frac{1}{2}\left(\rho + \frac{1}{\rho}\right); \quad \sqrt{F^2 - 1} = \frac{1}{2}\left(\rho - \frac{1}{\rho}\right). \tag{1.10}$$

This is a one-to-one correspondence and maps a neighborhood of the segment [-1, 1] in the F plane onto an exterior neighborhood of the unit circle, the circle itself corresponding to the segment. Since the $\rho - \lambda$ correspondence is one-to-one, the correspondence $F - \lambda$ is also one-to-one. The function $F(\rho)$ can be continued according to formula (1.10) into the interior of the circle and $F(\rho) = F(\rho^{-1})$. In view of the one-to-one correspondence, we have

$$\lambda(\rho) = \lambda(\rho^{-1}). \tag{1.11}$$

All functions that are regular in λ in the neighborhood of an arc of the spectrum, for example, $\varphi(x, \lambda)$, $\theta(x, \lambda)$, $\varphi'(x, \lambda)$, etc., can be considered as regular functions of ρ in the neighborhood of the unit circle. In the following, we will denote these functions for brevity by $\varphi(x, \rho)$, $\theta(x, \rho)$, $\varphi'(x, \rho)$, etc. It is obvious that the following equalities hold for them: $\varphi(x, \rho) = \varphi(x, \rho^{-1})$,

$\theta(x, \rho) = \theta(x, \rho^{-1})$, etc. Let us introduce the following notation:

$$m(\rho) \equiv \frac{\varphi'(\rho) - \theta(\rho)}{2\varphi(\rho)} + \frac{\rho^2 - 1}{2\rho\varphi(\rho)}, \tag{1.12}$$

$$\begin{aligned} \psi(x, \rho) &\equiv \theta(x, \rho) + m(\rho)\varphi(x, \rho), \\ \chi(x, \rho) &\equiv \psi(x, \rho)\rho^{-x}. \end{aligned} \tag{1.13}$$

It is easy to see that

$$\begin{aligned} m_1(\lambda) &= m(\rho), \quad m_2(\lambda) = m(\rho^{-1}), \\ \psi_1(x, \lambda) &= \psi(x, \rho), \quad \psi_2(x, \lambda) = \psi(x, \rho^{-1}), \\ \chi_1(x, \lambda) &= \chi(x, \rho), \quad \chi_2(x, \lambda) = \chi(x, \rho^{-1}). \end{aligned} \tag{1.14}$$

Everywhere in the preceding we can take combinations of functions which do not lose generativity at $\varphi(\rho) = 0$. In view of this, we will assume everywhere in the following that $\varphi(\rho) \neq 0$ when $|\rho| = 1$ without loss of generality. From the definition of $\psi(x, \rho)$ and Lemma 1, it follows that

$$\psi(x, \rho) \in \begin{cases} L_2(-\infty, 0) & (|\rho| > 1). \\ L_2(0, \infty) & (|\rho| < 1). \end{cases} \tag{1.15}$$

When $|\rho| = 1$, the function $\psi(x, \rho)$ is bounded over the whole of the axis. It is also a simple matter to calculate the Wronskian

$$W[\psi(x, \rho); \psi(x, \rho^{-1})] = m(\rho^{-1}) - m(\rho) = \frac{1 - \rho^2}{\rho\varphi(\rho)}. \tag{1.16}$$

Let us construct the function

$$\Gamma(x, \xi, \rho) \equiv -\frac{\rho\varphi(\rho)}{\rho^2 - 1}\begin{cases} \psi(x, \rho^{-1})\psi(\xi, \rho) & (x \geqslant \xi) \\ \psi(\xi, \rho^{-1})\psi(x, \rho) & (x < \xi) \end{cases} = -\frac{\rho^{-|x-\xi|+1}\varphi(\rho)}{\rho^2 - 1}\begin{cases} \chi(x, \rho^{-1})\chi(\xi, \rho) & (x \geqslant \xi), \\ \chi(\xi, \rho^{-1})\chi(x, \rho) & (x < \xi). \end{cases} \tag{1.17}$$

It is obvious that for $|\rho| > 1$, the function $\Gamma(x, \xi, \rho)$ is the Green's function of the operator $L_0 - \lambda(\rho)E$. In the following, we will require estimates of the derivatives of the numerator of $\Gamma(x, \xi, \rho)$. Let us prove the following lemma in this connnection.

LEMMA 2. *In a neighborhood of the circle $|\rho| = 1$, including the circle itself, the following estimates hold*

$$\left| \frac{\partial^k}{\partial\rho^k} \rho^{-|x-\xi|+1}\varphi(\rho)\chi(x, \rho)\chi(\xi, \rho^{-1}) \right| < \rho^{-|x-\xi|}[C(k + |x - \xi|)]^k, \tag{1.18}$$

where $k = 0, 1, 2, \ldots$ and C is a positive constant.

PROOF. Since $\varphi(\rho) \neq 0$ when $|\rho| = 1$, in a neighborhood of the unit circle we also have $\varphi(\rho) \neq 0$. Let the domain D' be the annulus $|2R|^{-1} < |\rho| < 2R$ with a cut along any ray, say, the ray $\arg\rho = \pi$, the constant R being chosen sufficiently close to unity so that $\varphi(\rho) \neq 0$ when $\rho \in D'$. We see from the definition of $\chi(x, \rho)$ that it is regular in D'. The same thing holds for the function $\Phi(\rho) \equiv \varphi(\rho)\chi(x, \rho)\chi(\xi, \rho^{-1})$. Moreover, $\Phi(\rho)$ is uniformly bounded in x and ξ with $x, \xi \in (-\infty, \infty)$. Let us now select a closed domain which is strictly inside D', for example, the domain

$$\overline{D}'': \left\{ \left(R + \frac{1}{2}\right)^{-1} \leqslant |\rho| \leqslant R + \frac{1}{2}, \quad -\frac{3}{4}\pi \leqslant \arg\rho \leqslant \frac{3}{4}\pi \right\}.$$

It is easy to see that the following estimates hold in $\bar{D}''$

$$\left|\frac{d^k}{d\rho^k}\Phi(\rho)\right| \leqslant k!\,N^k, \tag{1.19}$$

where $k = 0, 1, 2, \ldots, \rho \in \bar{D}''$, and N is a positive constant that depends only on R. It is clear that if we make the cut along the ray $\arg \rho = 0$, we will obtain analogous estimates in the domain

$$\bar{D}''' : \left\{\left(R+\frac{1}{2}\right)^{-1} \leqslant |\rho| \leqslant R+\frac{1}{2},\quad \frac{\pi}{4} \leqslant \arg\rho \leqslant \frac{7}{4}\pi\right\}.$$

It follows from this that estimates (1.19) are valid in the whole of the domain

$$\bar{D} : \left\{\left(R+\frac{1}{2}\right)^{-1} \leqslant |\rho| \leqslant R+\frac{1}{2}\right\}.$$

Let us now estimate the derivatives of the function $\rho^{-|x-\xi|+1}\Phi(\rho)$, namely,

$$\frac{\partial^k}{\partial\rho^k}[\rho^{-|x-\xi|+1}\Phi(\rho)] = \sum_{s=0}^{k} \frac{k(k-1)\ldots(k-s+1)}{s!}[\rho^{-|x-\xi|+1}]^{(k-s)} \times$$

$$\times [\Phi(\rho)]^{(s)} = \sum_{s=0}^{k} \frac{k\ldots(k-s+1)(-|x-\xi|+1)\ldots(-|x-\xi|-k+s+2)}{s!}\rho^{-|x-\xi|-k+s+1}[\Phi(\rho)]^{(s)}.$$

Making use of the well-known inequality

$$|x_1 x_2 \ldots x_n| \leqslant \left[\frac{|x_1|+|x_2|+\ldots+|x_n|}{n}\right]^n,$$

as well as estimates (1.19), we obtain

$$\left|\frac{\partial^k}{\partial\rho^k}[\rho^{-|x-\xi|+1}\Phi(\rho)]\right| < \sum_{s=0}^{k} N^s \rho^{-|x-\xi|-k+s+1}\left(\frac{k-1}{2}+\frac{s}{k}|x-\xi|\right)^k < (k+|x-\xi|)^k \rho^{-|x-\xi|-k+1}[N\rho]^k.$$

The lemma now follows directly from this.

Let us now prove an auxiliary assertion which will be used below.

LEMMA 3. Let the function $\Omega(z)$ be regular within a domain D_1 and continuous together with all of its derivatives right up to the boundary l of D_1. Then, the following equalities hold for any z belonging to D_1. and any z_0 belonging to l:

$$\frac{d^k}{dz^k}\frac{\Omega(z)}{z-z_0} = \frac{(-1)^k k!\,\Omega(z_0)}{(z-z_0)^{k+1}} + \Omega_{k+1}^{z_0}(z), \tag{1.20}$$

where $\Omega_{k+1}^{z_0}(z)$ is a function regular in D_1 and continuous right up to the boundary satisfying the inequality

$$|\Omega_{k+1}^{z_0}(z)| < 2^k \max_{z\in\bar{D}_1}|\Omega^{(k+1)}(z)|. \tag{1.21}$$

PROOF. Let us write Taylor's formula for the function $\Omega(z)$ and divide it by $z - z_0$:

$$\frac{\Omega(z)}{z-z_0} = \frac{\Omega(z_0)}{z-z_0} + \frac{\Omega'(z_0)}{1!} + \ldots + \frac{\Omega^{(k)}(z_0)(z-z_0)^{k-1}}{k!} + \frac{R_k(z)}{z-z_0}.$$

Let us express the remainder as an integral, namely,

$$R_k(z) = \frac{1}{k!}\int_{z_0}^{z} \Omega^{(k+1)}(t)(z-t)\,dt.$$

Let us differentiate the resulting inequality k times and let us write

$$\Omega_{k+1}^{z_0}(z) \equiv \frac{d^k}{dz^k}\frac{R_k(z)}{z-z_0}.$$

We then obtain formula (1.20), namely,

$$\frac{d^k}{dz^k}\frac{\Omega(z)}{z-z_0} = (-1)^k \frac{k!\,\Omega(0)}{(z-z_0)^{k+1}} + \Omega_{k+1}^{z_0}(z).$$

We must now derive inequality (1.21). To do this, we make use of Leibnitz's formula

$$\frac{d^k}{dz^k}\frac{R_k(z)}{z-z_0} = \sum_{s=0}^{k} \frac{s!\,(-1)^s R_k^{(k-s)}(z)}{(z-z_0)^{s+1}} C_k^s. \tag{1.22}$$

The derivatives of $R_k(z)$ can be easily found to be

$$\frac{d^m}{dz^m} R_k(z) = \frac{1}{(k-m)!}\int_{z_0}^{z} \Omega^{(k+1)}(t)(z-t)^{k-m}\,dt,$$

from which inequality (1.21) follows immediately.

Let us apply Lemma 3 to $\Gamma(x, \xi, \rho)$. Let us take D_1 to be the semi-annulus $1 < |\rho| < R + 1/2$, $\operatorname{Re}\rho > 0$. If we set $\chi(x) \equiv \chi(x, 1)$, then the derivatives of $\Gamma(x, \xi, \rho)$ can be represented as

$$\frac{\partial^k}{\partial\rho^k}\Gamma(x, \xi, \rho) = \frac{\varphi(1)\chi(x)\chi(\xi)}{2(\rho-1)^{k+1}} + \Gamma_{k+1}^1(x, \xi, \rho). \tag{1.23}$$

Lemmas 2 and 3 allow us to assert that the function $\Gamma_{k+1}^1(x, \xi, \rho)$ is regular in D_1, continuous right up to the boundary, and satisfies the inequality

$$|\Gamma_{k+1}^1(x, \xi, \rho| < A^k \rho^{-|x-\xi|}(k+|x-\xi|)^k < A^k(k+|x-\xi|)^k \tag{1.24}$$

($k = 0, 1, 2, \ldots$, and A is a positive constant which depends only on D_1). It is obvious that formulas analogous to (1.23) and (1.24) also hold in the vicinity of the point $\rho = -1$.

§ 2. Investigations of the Perturbed Operator

Let us reduce the investigation of the differential operator to the investigation of an integral operator and let us study the resolvent of the latter.

Let $q_1(x)$ and $q_2(x)$ be real functions defined over the whole of the axis and let the function $q(x) = q_1(x) + iq_2(x)$ satisfy the following conditions:

$$|q(x)x^k| < f_k f(x), \quad k = 0, 1, 2, \ldots, \tag{2.1}$$

where $f(x) \in L(-\infty, \infty)$ with $f(x) > 0$ and the f_k are positive constants. We will investigate the operator L defined by

$$Lu \equiv L_0 u + q(x)u = -u'' + p(x) + q(x)u, \quad u \in L_2(-\infty, \infty).$$

It can be easily seen on the basis of general operator-theoretic considerations that the continuous spectra of L and L_0 coincide, while the eigenvalues of the operator L, if they exist, can only accumulate toward points of the continuous spectrum. We will study the spectrum of the operator L in the vicinity of the arbitrary arc of the continuous spectrum chosen earlier and we will show that under some conditions the eigenvalues of L do not have accumulation points on the arc.

Let us select a domain Λ_1 in the vicinity of the chosen arc of the spectrum in such a manner that $\rho(\lambda)$ is in D_1: $\{1 \le |\rho| \le R + 1/2, \operatorname{Re} \rho > 0\}$ for $\lambda \in \Lambda_1$. Let u with $u \in L_2(-\infty, \infty)$ be a solution of the equation

$$Lu = \lambda u, \quad \lambda \in \Lambda_1. \tag{2.2}$$

Multiplying both sides of this equation $\Gamma[x, \xi, \rho(\lambda)]$ and integrating over ξ, we obtain the integral equation

$$u(x) = \int_{-\infty}^{\infty} \Gamma[x, \xi, \rho(\lambda)]\, q(\xi)\, u(\xi)\, d\xi.$$

Let us write down an auxiliary equation*

$$v(x) = \mu \int \Gamma[x, \xi, \rho(\lambda)]\, q(\xi)\, v(\xi)\, d\xi. \tag{2.3}$$

It is obvious that the values of λ for which Eq. (2.3) has an eigenvalue $\mu = 1$ are the eigenvalues of the operator L. Equation (2.3) is a homogeneous Fredholm equation with a Hilbert–Schmidt kernel for all ρ belonging to D_1. The kernel of its resolvent has the usual representation

$$r(x, \xi, \rho, \mu) = \frac{d(x, \xi, \rho, \mu)}{d(\rho, \mu)},$$

where

$$d(x, \xi, \rho, \mu) \equiv -\Gamma(x, \xi, \rho)\, q(\xi) + \sum_{n=1}^{\infty} \frac{(-\mu)^n}{n!} \int_{E_n} \Gamma_\rho \begin{pmatrix} x, & s_1, & \ldots & s_n \\ \xi, & s_1, & \ldots & s_n \end{pmatrix} \prod_{i=1}^{n} q(s_i)\, ds_i,$$

$$d(\rho, \mu) \equiv 1 + \sum_{n=1}^{\infty} \frac{(-\mu)^n}{n!} \int_{E_n} \Gamma_\rho \begin{pmatrix} s_1, & \ldots & s_n \\ s_1, & \ldots & s_n \end{pmatrix} \prod_{i=1}^{n} q(s_i)\, ds_i, \tag{2.4}$$

$$\Gamma_\rho \begin{pmatrix} s_1, & \ldots & s_n \\ s_1, & \ldots & s_n \end{pmatrix} \equiv \det_n \{\Gamma(s_i, s_j, \rho)\}.$$

* Here and in the following, an integral without explicit limits denotes an integration over the whole of the axis $(-\infty, \infty)$ and an integral with a subscript E_n denotes integration over the whole of the n-dimensional space.

Assuming that $\mu = 1$, we obtain

$$R(x, \xi, \rho) = \frac{D(x, \xi, \rho)}{D(\rho)}.$$

Here, we have

$$D(x, \xi, \rho) \equiv d(x, \xi, \rho, 1),$$

$$D(\rho) \equiv d(\rho, 1) = 1 + \sum_{n=1}^{\infty} \frac{(-1)^n}{n!} \int_{E_n} \Gamma_\rho \begin{pmatrix} s_1, & \ldots & s_n \\ s_1, & \ldots & s_n \end{pmatrix} \prod_{l=1}^{n} q(s_l)\, ds_l. \tag{2.5}$$

The eigenvalues of L coincide with the zeros of the function $D(\rho)$. In the following, we will be occupied with the study of the distribution of the zeros of $D(\rho)$.

Let us first formulate a simple proposition.*

LEMMA 4. Let $T = \{t_{ij}\}$ be quadratic matrices of order n and let $p_{ij} = p_i r$. Then, we have

$$\det(T + P) = \det T + \sum_{i, j=1}^{n} p_i r_j T_{ij}, \tag{2.6}$$

where T_{ij} is the cofactor of t_{ij}.

Let us now write down the k-th derivative of $D(\rho)$. It follows from formula (2.5) that

$$\frac{d^k}{d\rho^k} D(\rho) = \sum_{n=1}^{\infty} \frac{(-1)^n}{n!} \sum_{k_1+k_2+\ldots+k_n=k} \frac{k!}{k_1! k_2! \ldots k_n!} \int_{E_n} \det{}_n \{\Gamma^{(k_i)}(s_i, s_j, \rho)\} \prod_{l=1}^{n} q(s_l)\, ds_l. \tag{2.7}$$

Here, we have

$$\Gamma^{(k_i)}(s_i, s_j, \rho) \equiv \frac{\partial^{k_i}}{\partial \rho^{k_i}} \Gamma(s_i, s_j; \rho).$$

We have to show that the series in (2.7) converges uniformly in ρ when $\rho \in \overline{D}_1$ and to obtain the estimate of the derivative $D^{(k)}(\rho)$. Let us first of all evaluate the integral

$$\int_{E_n} \det \{\Gamma^{(k_i)}(s_i, s_j, \rho)\} \prod_{l=1}^{n} q(s_l)\, ds_l.$$

We will make use of Lemma 4 in evaluating this integral. Let us introduce the abbreviations

$$a_i \equiv \frac{\varphi(1)(-1)^{k_i} k_i!}{2}, \quad r_l \equiv \chi(s_l),$$
$$t_{ij} \equiv \Gamma^1_{k_i+1}(s_i, s_j, \rho), \quad p_i \equiv \frac{a_l r_l}{(\rho-1)^{k_l+1}}. \tag{2.8}$$

* This lemma is a special case of a result obtained by Tamarkin [3]. A proof which is simpler than the one given in [3] can be found in [4].

Applying formula (1.23), as well as Lemma 4, we obtain the following representation

$$\det_n \{\Gamma^{(k_i)}(s_i, s_j, \rho)\} = \det T^k + \sum_{i,j=1}^{n} \frac{a_i r_i r_j}{(\rho-1)^{k_i+1}} T_{ij}^k, \tag{2.9}$$

where $T^k \equiv \{t_{ij}\}$ and T_{ij}^k is the cofactor of t_{ij} in T^k. Let us take $k_i = 0$ in formula (2.9), so that we have

$$\det_n \{\Gamma(s_i, s_j, \rho)\} = \det T^0 + \frac{\varphi(1)}{2(\rho-1)} \sum_{i,j=1}^{n} \chi(s_i)\chi(s_j) T_{ij}^0. \tag{2.10}$$

If we substitute this equality into formula (2.5), we see that as the point $\rho = 1$ is approached, the function $D(\rho)$ increases as $(\rho - 1)^{-1}$. Therefore, it is more convenient to deal with the function $\Delta(\rho) = (\rho - 1)D(\rho)$. Let us write down the explicit expression for $\Delta(\rho)$

$$\Delta(\rho) = (\rho-1)\left(1 + \sum_{n=1}^{\infty} \frac{(-1)^n}{n!} \int_{E_n} \det T^0 \prod_{l=1}^{n} q(s_l)\, ds_l\right) + \frac{\varphi(1)}{2} \sum_{n=1}^{\infty} \frac{(-1)^n}{n!} \sum_{i,j=1}^{n} \int_{E_n} \chi(s_i)\chi(s_j) T_{ij}^0 \prod_{l=1}^{n} q(s_l)\, ds_l. \tag{2.11}$$

If we calculate the derivative $D^{(k)}(\rho)$, we will obtain a series in powers of $1/(\rho - 1)$. It is clear that the derivative $\Delta^{(k)}(\rho)$ will be the coefficient of $(\rho - 1)^{-1}$ in this expression. Let us write down the expression for this derivative. The prime on the summation sign will denote that we are only considering the terms in which $k_i = 0$. Making use of formulas (2.7) and (2.9), we obtain

$$\Delta^{(k)}(\rho) = \sum_{k=1}^{\infty} \frac{(-1)^n}{n!} \sum_{k_1+\dots+k_n=k} \frac{k!}{k_1! \dots k_n!} \int_{E_n} \left\{(\rho-1)\det T^k + \sum_{ij=1}^{n}{}' a_i r_i r_j T_{ij}^k\right\} \prod_{l=1}^{n} q(s_l)\, ds_l. \tag{2.12}$$

We now have to evaluate the integrals appearing in formula (2.12). Let us investigate the integrand. First of all, we prove the following lemma.

LEMMA 5. *The following estimate holds:*

$$\left|\det T^k \prod_{i=1}^{n} q(s_i)\right| < [nf_0]^{\frac{n}{2}} B^k \prod_{i=1}^{n} f(s_i)\, \varkappa_i, \tag{2.13}$$

where

$$\varkappa_i \equiv (k_i+1)^{k_i+1}\sqrt{f_0} + 2^{k_i+1}\sqrt{f_{2k_i+2}} \tag{2.14}$$

and the constants f_{2k_i+2}, f_0, and the function f are the same as those appearing in conditions (2.1).

PROOF. According to formula (1.24), we have

$$|t_{ij}| < A^{k_i+1}[(k_i+1) + |s_i - s_j|]^{k_i+1} \leqslant (2A)^{k_i+1}[(k_i+1)^{k_i+1} + |s_i - s_j|^{k_i+1}] \leqslant$$
$$\leqslant (2A)^{k_i+1}[(k_i+1)^{k_i+1} + 2^{k_i}(|s_i|^{k_i+1} + |s_j|^{k_i+1})]. \tag{2.15}$$

Condition (2.1) can be rewritten as

$$\left|\frac{q(x)}{f(x)}\right| < \frac{f_k}{|x|^k}.$$

Making use of this inequality and formula (2.15), we obtain

$$|t_{ij}\sqrt{q(s_i)q(s_j)}| < B^{k_i+1}\sqrt{f(s_i)f(s_j)}\left[(k_i+1)^{k_i+1}\sqrt{\frac{|q(s_i)q(s_j)|}{f(s_i)f(s_j)}} + 2^{k_i}\left(\sqrt{\frac{|s_i|^{2k_i+2}|q(s_i)|}{f(s_i)}}\sqrt{\frac{|q(s_j)|}{f(s_j)}} + \right.\right.$$
$$\left.\left. + \sqrt{\frac{|s_j|^{2k_i+2}|q(s_j)|}{f(s_j)}}\sqrt{\frac{|q(s_i)|}{f(s_i)}}\right)\right] < F_i\sqrt{f(s_j)}, \tag{2.16}$$

where

$$B = 2A, \quad F_i = B^{k_i+1}\sqrt{f(s_i)f_0}\,\varkappa_i.$$

Let us rewrite inequality (2.16) as

$$\left|\frac{t_{ij}\sqrt{q(s_i)q(s_j)}}{F_i}\right| < \sqrt{f(s_j)}. \tag{2.17}$$

We can now easily evaluate the determinant

$$\det\{t_{ij}\}\prod_{l=1}^{n} q(s_l) = \det\{t_{ij}\sqrt{q(s_i)q(s_j)}\} = \det\left\{\frac{t_{ij}\sqrt{q(s_i)q(s_j)}}{F_i}\right\}\prod_{l=1}^{n} F_l.$$

From Hadamard's inequality and formula (2.17) we obtain

$$\left|\det\left\{\frac{t_{ij}\sqrt{q(s_i)q(s_j)}}{F_i}\right\}\right|^2 \leqslant \prod_{j=1}^{n}\sum_{i=1}^{n}\left|\frac{t_{ij}\sqrt{q(s_i)q(s_j)}}{F_i}\right|^2 < n^n\prod_{j=1}^{n} f(s_j).$$

This leads directly to estimate (2.13).

LEMMA 6. *The following inequality holds*

$$\left|\sum_{i,j=1}^{n}{}' T_{ij}^k a_i r_i r_j \prod_{l=1}^{n} q(s_l)\right| < [nf_0]^{\frac{n}{2}} B_1^n \prod_{i=1}^{n} f(s_i)\varkappa_i, \tag{2.18}$$

where B_1 is a positive constant.

PROOF. The function $r_i = \chi(s_i)$ is periodic in x and, consequently, i is bounded. Therefore, we can choose B_1 such that

$$|a_i r_i r_j| < B_1 < B_1^{k_i+1}(k_i+1)!\,[(k_i+1)+|s_i - s_j|]^{k_i+1}.$$

Clearly, this is a crude estimate, but it is a convenient one for our proof. Let us note that $\sum_{j=1}^{n} a_i r_i r_j T_{ij}^k$ is nothing else but the determinant $\det\{t_{lm}\}$ in which the elements of the i-th row, namely, t_{ij} with $j = 1, 2, \ldots, n$, are replaced by $a_i r_i r_j$. The proof now becomes completely analogous to that of Lemma 5.

Let G_k^n denote the number of ways in which n nonnegative integers can be selected to add up to k. It can be shown that

$$G_k^n < Ce^{n+k}. \tag{2.19}$$

Let us also introduce the abbreviation $\gamma \equiv \int f(x)\,dx$, where $f(x)$ is the same function as in (2.1). The following inequalities now follow directly from (2.13) and (2.18):

$$\left| \sum_{\Sigma k_i = k} \frac{k!}{k_1! \dots k_n!} \int_{E_n} \det T^k \prod_{l=1}^{n} q(s_l)\, ds_l \right| < [f_0 \gamma^2 n]^{\frac{n}{2}} B^k G_k^n k! \max_{\Sigma k_i = k} \prod_{i=1}^{n} \frac{\varkappa_i}{k_i!}, \tag{2.20}$$

$$\left| \sum_{\sum_{m=1}^{n} k_m = k} \frac{k!}{k_1! \dots k_n!} \int_{E_n} {\sum_{i,j=1}^{n}}' a_i r_i r_j T_{ij}^k \prod_{l=1}^{n} q(s_l)\, ds_l \right| < [n f_0 \gamma^2]^{\frac{n}{2}} n B_1^k k! G_k^n \max_{\Sigma k_i = k_n} \prod_{i=1}^{n} \frac{\varkappa_i}{k_i!}. \tag{2.21}$$

REMARK. It can be easily seen that $\prod_{i=1}^{n} \frac{\varkappa_i}{k_i}$ has the following rough estimate:

$$\left| \prod_{i=1}^{n} \frac{\varkappa_i}{k_i!} \right| < C^n,$$

where C is a positive constant. This estimate allows us to assert that the series in formula (2.12) converge uniformly in domain D_1 for any value of k. It follows from this that $\Delta(\rho)$ is regular in D_1 and is continuous together with all of its derivatives right up to the boundary of this domain.

We will assume in the following that stronger conditions are imposed on the perturbing potential.

LEMMA 7. Under the conditions

$$|q(x)| < C \exp(-\delta |x|^{\alpha}), \quad (C > 0,\ \delta > 0,\ 0 < \alpha < 1), \tag{2.22}$$

$\Delta^{(k)}(\rho)$ satisfy the inequalities

$$|\Delta^{(k)}(\rho)| < k^{\frac{k}{\alpha}} B_2^k, \quad (B_2 > 0). \tag{2.23}$$

PROOF. Let us take $f(x) = (1 + x^2)^{-1}$ in formula (2.1). Then, it is obvious that we can take

$$f_k = C^k (k+2)^{\frac{k+2}{\alpha}}, \quad k = 2, 3, \dots \tag{2.24}$$

Indeed, we have

$$\sup_{x \in (-\infty, \infty)} |q(x)|\, |x|^k (1 + x^2) < C \sup \exp(-\delta |x|^{\alpha}) |x|^{k+2} =$$

$$= C \sup_{x \in (0, \infty)} \exp[-\delta x^{\alpha} + (k+2) \ln x] = C^k (k+2)^{\frac{k+2}{\alpha}}.$$

Let us now obtain a more accurate estimate of $\prod_{i=1}^{n} \frac{\varkappa_i}{k_i!}$. We will make use of the well-known inequality

$$\frac{1}{m!} < \frac{e^m}{m^m} \ (m > 0).$$

It follows from formulas (2.14) and (2.24) that

$$\left| \prod_{i=1}^{n} \frac{\varkappa_i}{k_i!} \right| < \prod_{i=1}^{n} \left[\frac{(k_i+1)^{k_i+1}}{k_i!} \sqrt{f_0} + C^k \frac{(2k_i+4)^{(2k_i+4)/\alpha}}{k_i!} \right] < C^{k+r} \prod_{i=1}^{n} k_i^{k_i/\alpha - k_i}.$$

Finally, we have

$$\left| \prod_{i=1}^{n} \frac{\varkappa_i}{k_i!} \right| < C^{k+n} k^{\frac{k}{\alpha} - k}. \tag{2.25}$$

Here C_1 is a positive constant.

It is now relatively easy to estimate the derivatives of $\Delta(\rho)$ defined by (2.12) and we have

$$|\Delta^{(k)}(\rho)| < k^{\frac{k}{\alpha} - k} C^k k! \sum_{n=1}^{\infty} \frac{e^n [n f_0 \gamma^2]^{\frac{n}{2}} n}{n!}.$$

The assertion of the lemma now follows directly from this inequality.

Pavlov [2] has shown that if inequalities of the form (2.23) hold for the derivatives of a function which is regular in the upper half-plane and which, together with all of its derivatives, is continuous right up to the boundary, then the following assertions are valid.

1. If $0 < \alpha < 1/2$, then the set of points at which the zeros of this function accumulate is bounded, closed, of linear measure zero, and satisfies the condition

$$\sum_{\nu} |l_\nu|^{\frac{1-2\alpha}{1-\alpha}} < \infty.$$

Here, $|l_\nu|$ is the length of the interval l_ν of contiguity with the zero-accumulation set and the summation is extended over all intervals l_ν.

2. When $\alpha > 1/2$, the accumulation points are completely absent.

These assertions can be easily extended to the case of the exterior of a unit circle and, consequently, they are valid for our domain D_1.

The domain D_1 encompasses the right half of the circle $|\rho| = 1$. We can also examine the neighborhood of the left half of the unit circle in the same way. In particular, this leads to the following assertion.

THEOREM 1. *Let L be an operator in $L_2(-\infty, \infty)$ of the form*

$$Lu = -u'' + p(x)u + q(x)u,$$

where p(x) is a periodic complex function. If q(x) satisfies the condition

$$|q(x)|<C\exp\left(-\delta|x|^{\frac{1}{2}}\right),$$

then the set of eigenvalues of L cannot have accumulation points at a finite distance from the origin.

§ 3. Investigation of the Spectrum under the Condition $\int|q(x)|e^{\delta|x|}dx<\infty$

We will show in this section that analogous conclusions concerning the spectrum of L can be reached much more simply if we subject the perturbing potential to the stronger requirement

$$|q(x)|<C\exp(-\delta|x|)\ (\delta>0). \tag{3.1}$$

It will be shown that under condition (3.1), the neighborhood of the unit circle $|\rho|=1$ does not contain more than a finite number of points ρ_n at which Eq. (2.3) has an eigenvalue $\mu=1$.

Let us first of all slightly modify Eq. (2.3). Let us multiply both sides of it by $\sqrt{q(x)}$ and let us introduce the abbreviations

$$\begin{aligned}&\sqrt{q(x)}\,u(x)\equiv w(x),\\&\sqrt{q(x)q(\xi)}\,\Gamma(x,\xi,\rho)\equiv K(x,\xi,\rho).\end{aligned} \tag{3.2}$$

We can now rewrite (2.3) as

$$w(x)=\mu\int K(x,\xi,\rho)\,w(\xi)\,d\xi. \tag{3.3}$$

It should be noted that if for some ρ_0 there exists a solution $u(x,\rho_0)$ of Eq. (2.3) with $u(x,\rho_0)$ $\in L_2(-\infty,\infty)$, then all the more we have $w(x,\rho_0)$ is a solution of Eq. (3.3). Therefore, any eigenvalue of an operator with kernel $\Gamma(x,\xi,\rho)\,q(\xi)$ is an eigenvalue of an operator with kernel $K(x,\xi,\rho)$.

We will carry out our investigation in the domain

$$D_2:\left\{r^{-1}<|\rho|<r,\ -\frac{2\pi}{3}<\arg\rho<\frac{2\pi}{3}\right\},$$

choosing r sufficiently close to unity so that we have

$$\ln r<\frac{\delta}{3} \tag{3.4}$$

[here δ has the same meaning as in formula (3.1)].

The kernel $K(x,\xi,\rho)$ is a meromorphic function of ρ in D_2. Kernels of this type have been studied by Tamarkin [3]. Let us first of all establish some of the properties of $K(x,\xi,\rho)$ so that we can use Tamarkin's results.

1. $K(x,\xi,\rho)$ is regular in D with the exception of the point $\rho=1$ at which it has a simple pole for all x and ξ; the following inequality is satisfied by $K(x,\xi,\rho)$:

$$|K(x,\xi,\rho)|<C\frac{\exp\left[-\frac{\delta}{6}(|x|+|\xi|)\right]}{\rho-1}, \tag{3.5}$$

which follows directly from formulas (1.17) and (3.1)-(3.4). Inequality (3.5) allows us to make the following assertion.

2. In any open subdomain $D_3 \subset D_2$ which does not contain the point $\rho = 1$, the kernel $K(x, \xi, \rho)$ is a Carleman kernel, while in any closed subdomain $D_4 \subset D_3$, we have

$$\int |K(x, \xi, \rho)|^2 d\xi \leqslant K_0(x),$$
$$\int |K(x, \xi, \rho)|^2 dx \leqslant K_0(\xi),$$

where $K_0(x)$ is a positive function which depends only on D_4 and which is integrable in $(-\infty, \infty)$.

The kernel $K(x, \xi, \rho)$ also possesses the following property.

3. The coefficient of $(\rho - 1)^{-1}$ in the expansion of $K(x, \xi, \rho)$ in the vicinity of the hole $\rho = 1$ can be represented as a product of a function of x only by a function of ξ only, these factors belonging to $L_2(-\infty, \infty)$.

In fact, assuming that $k = 0$ in formula (1.23), we obtain

$$K(x, \xi, \rho) = \frac{\varphi(1)\left[\sqrt{q(x)}\,\chi(x)\right]\left[\sqrt{q(\xi)}\,\chi(\xi)\right]}{\rho - 1} + \\ + \Gamma_1^1(x, \xi, \rho)\sqrt{q(x)\,q(\xi)}. \tag{3.6}$$

The quadratic integrability of the function follows from condition (3.1).

Tamarkin [3] has shown that if conditions 1-3 are satisfied, then the resolvent of $K(x, \xi, \rho)$ is meromorphic with respect to ρ in D_2. We can carry out an analogous investigation in the neighborhood of the point $\rho = -1$.

From the fact that the resolvent is meromorphic in the neighborhood of the unit circle, it follows that on the circle there are no points of accumulation of values of ρ such that $K(x, \xi, \rho)$ has an eigenvalue $\mu = 1$. Consequently, there can be no points of accumulation of the eigenvalues of L on an arc of the continuous spectrum.

§4. Proof That There Are No Eigenvalues on the Continuous Spectrum

It has been shown in [5] that under the condition

$$\int |q(x)|(1 + |x|)\,dx < \infty \tag{4.1}$$

there exist solutions of Eq. (2.2) of the form

$$f_i(x, \lambda) = \psi_i(x, \lambda) - \int_x^\infty G(x, \xi, \lambda)\,q(\xi)\,\psi_i(\xi, \lambda)\,d\xi, \tag{4.2}$$

where $i = 1, 2$, the $\psi_i(x, \lambda)$ are functions defined by (1.4), and $G(x, \xi, \lambda)$ is the Cauchy function of Eq. (2.2). If λ is a point of the continuous spectrum of the operator L, then $G(x, \xi, \lambda)$ satisfies the inequality (see [5])

$$|G(x, \xi, \lambda)| \leqslant C(1 + |x - \xi|),\ C > 0. \tag{4.3}$$

If, as was mentioned in Section 1, λ belongs to the spectrum of the operator L_0, then we can write

$$\psi_1(x, \lambda) = e^{itx}\chi_1(x, \lambda),$$
$$\psi_2(x, \lambda) = e^{-itx}\chi_2(x, \lambda), \tag{4.4}$$
$$t \in [0, 2\pi).$$

When we have $t \in (0, 2\pi)$, $t \neq \pi$, the function ψ_1 and ψ_2 are linearly independent. When $t = 0$ and $t = \pi$, we have $\psi_1(x, \lambda) = \psi_2(x, \lambda)$: these points correspond to the ends of the arc of the spectrum. It can be seen from condition (4.1) and formulas (4.3) and (4.4) that if λ belongs to the spectrum

$$\int_x^\infty G(x, \xi, \lambda) q(\xi) \psi_i(\xi, \lambda) \, d\xi = o(1) \tag{4.5}$$

as $x \to \infty$.

Let us consider the following two cases separately: 1) λ is an inner point of the continuous spectrum; 2) λ is an edge point of one of the arcs forming the continuous spectrum.

LEMMA 8. *Under condition (4.1) none of the inner points of the continuous spectrum of the operator L_0 is an eigenvalue.*

PROOF. Let λ_1 be an inner point of the spectrum. Let us assume that there exists a solution $f(x, \lambda_1)$ of Eq. (2.2) with $\lambda = \lambda_1$ and that $f(x, \lambda_1) \in L_2(-\infty, \infty)$. It can be easily shown that if λ_1 is an inner point of the spectrum, then the functions $f_1(x, \lambda_1)$ and $f_2(x, \lambda_1)$ are linearly independent and we have

$$f(x, \lambda_1) = f_1(x, \lambda_1) + M f_2(x, \lambda_1).$$

It follows from formulas (4.5) and (4.4) that as $x \to \infty$ we have

$$f(x, \lambda_1) = \psi_1(x, \lambda_1) + M\psi_2(x, \lambda_1) + o(1) = \\ = e^{itx}\chi_1(x, \lambda_1) + Me^{-itx}\chi_2(x, \lambda_1) + o(1).$$

For $f(x, \lambda_1)$ to belong to $L_2(-\infty, \infty)$, it is necessary that we have

$$e^{2itx}\chi_1(x, \lambda_1) + M\chi_2(x, \lambda_1) \in L_2(a, \infty), \tag{4.6}$$

where a is a positive constant. It is clear that condition (4.6) cannot be satisfied when $t \neq \pi$.

LEMMA 9. *Under the condition*

$$\int |q(x)| (1 + x^2) \, dx < \infty \tag{4.7}$$

the edge points of the arcs of the continuous spectrum of the operator L cannot be eigenvalues.

PROOF. Let $\lambda_0(1)$ be an edge point of an arc of the spectrum. We then have $m_1(\lambda_0) = m_1(\lambda_0)$ and

$$\psi_1(x, \lambda_0) = \psi_2(x, \lambda_0) = \psi(x, 1) = \chi_1(x, \lambda_0) = \chi_2(x, \lambda_0) = \chi(x, 1).$$

It follows from this that $f_1(x, \lambda_0) = f_2(x, \lambda_0)$. We must find a solution of Eq. (2.2) that is linearly independent of $f_1(x, \lambda_0)$. Let us make use of the function $\varphi(x, \lambda)$ for this purpose. Formula (1.4) directly leads to

$$\varphi(x, \lambda) = \frac{\psi_1(x, \lambda) - \psi_2(x, \lambda)}{m_1(\lambda) - m_2(\lambda)} = \frac{[\psi(x, \rho) - \psi(x, \rho^{-1})]\rho}{(\rho - 1)(\rho + 1)}.$$

Eliminating the indeterminacy for $\rho \to 1$, we obtain

$$\lim_{\rho \to 1} \frac{[\exp(x \ln \rho)\chi(x, \rho) - \exp(-x \ln \rho)\chi(x, \rho^{-1})]\rho}{(\rho - 1)(\rho + 1)} = 2x\chi(x, 1) + 2\dot{\chi}_\rho(x, 1).$$

The solution Eq. (1.1) we have obtained, namely,

$$\varphi(x, \lambda_0) = 2x\chi(x, 1) + 2\dot{\chi}_\rho(x, 1)$$

is linearly independent of $\chi(x, 1)$. It is obvious that the function $\dot{\chi}_\rho(x, 1)$, as well as $\chi(x, 1)$, is periodic in x with unit period and, consequently, it is bounded in x. Let us construct the solution of Eq. (2.2) as

$$\tilde{\varphi}(x) \equiv \varphi(x, \lambda_0) - \int_x^\infty G(x, \xi, \lambda_0) q(\xi) \varphi(\xi, \lambda_0)\, d\xi.$$

Condition (4.7) and inequality (4.3) guarantee the convergence of the integral and we have

$$\tilde{\varphi}(x) - \varphi(x, \lambda_0) = o(1) \quad \text{as} \quad x \to \infty.$$

Let us now assume that for some M_1 we have

$$\tilde{\varphi}(x) + M_1 f_1(x, \lambda_0) \in L_2(-\infty, \infty).$$

The necessary condition for this is

$$\chi(x, 1)(2x + M_1) + 2\dot{\chi}_\rho(x, 1) \in L_2(a, \infty).$$

It is obvious that this is an impossible condition. Let us formulate this result in the form of a theorem.

THEOREM 2. *Under the condition*

$$\int |q(x)|(1 + |x|)\, dx < \infty$$

not one of the inner points of the continuous spectrum of the operator L can be an eigenvalue.

Under the condition

$$\int |q(x)|(1 + x^2)\, dx < \infty$$

the end points of the arcs of the continuous spectrum cannot be eigenvalues.

The author would like to express his gratitude to M. Sh. Birman and B. S. Pavlov for their supervision of the work.

LITERATURE CITED

1. F. S. Rofe-Beketov, "The spectra of nonself-adjoint differential operators with periodic coefficients," Dokl. Akad. Nauk SSSR, Vol. 152, No. 6 (1963).
2. B. S. Pavlov, "The nonself-adjoint Schroedinger operator," in Topics in Mathematical Physics, Vol. 1, Consultants Bureau, New York (1967).

3. J. D. Tamarkin, "On Fredholm's integral equations whose kernels are analytic in a parameter," Am. Math., Vol. 28, No. 2 (1927).

4. V. A. Zheludev, "Eigenvalues of the perturbed Schroedinger operator with a periodic potential," in Topics in Mathematical Physics, Vol. 2, Consultants Bureau, New York (1968).

5. F. S. Rofe-Beketov, "The criterion for the finiteness of the number of discrete levels introduced into lacunas of the continuous spectrum by a perturbation of a periodic potential," Dokl. Akad. Nauk SSSR, Vol. 156, No. 3 (1964).

6. E. C. Titchmarsh, Eigenfunction Expansions Associated with Second-Order Differential Equations, Vol. 2, Oxford University Press (1962).

7. V. A. Tkachenko, "The spectral analysis of the one-dimensional Schroedinger operator with a periodic potential," Dokl. Akad. Nauk SSSR, Vol. 155, No. 2 (1964).

THE DISCRETE SPECTRA OF THE DIRAC AND PAULI OPERATORS

O. I. Kurbenin

The aim of the present article is the investigation of the spectra of the Dirac and Pauli operators by operator-theoretic methods. This investigation is based on the estimation of the quadratic forms of the above operators by means of the quadratic forms of operators with well-known spectra. We may count the Schroedinger operator among the latter. Estimates of this type extended to both sides allow us, in some cases, to establish the criteria for the total multiplicity of the spectrum of the perturbed operator to be finite or infinite in that part of the axis which in the case of the unperturbed operator is a gap (i.e., free from the spectrum). Qualitative conclusions on the character of the spectrum and quantitative estimates of the total multiplicity of the spectrum follow from the well-known properties of the Schroedinger operator. In some cases, these conclusions refer to a family of operators obtained through the introduction of a parameter h which plays the part of Planck's constant in quantum mechanics. Skachek [6] has recently obtained the lower bound to the number of eigenvalues of the Dirac operator. By contrast with the results of [6], our estimates do not contain terms involving the derivative of the potential. Some results concerning the spectrum of the Pauli operator have been obtained by Glazman [4].

The spectral properties of the Schroedinger operator that we need have been studied in detail by Birman [1], who also obtained some results concerning the spectrum of the Dirac operator and described the method used by us below. The results and theorems of [1] will be frequently used in our proofs. For brevity, references to these theorems will be by the number used to identify them in [1]. The notation of [1] has been used throughout and is summarized in Section 1.

Next, Section 2 contains results on the nature of the spectrum of the Dirac operator in the interval $(-1, 1)$ for the case of a spherically symmetric bounded scalar potential q(r). The spectrum of the three-dimensional Dirac operator in $(-1, 1)$ is studied in Section 3. The results of Section 4 refer to the discrete part of the Pauli-operator spectrum.

§ 1. Auxiliary Information

1. We will make use of the following concepts and notation introduced in [1]:

1) p(x) will be said to be a radial function if $p(x) = p(|x|)$ and an almost radial function if $p(x) \geq 0$ and $\max_{|x|=r} p(x) \leqslant C \min_{|x|=r} p(x)$, where $C > 0$;

2) function p(x) defined and continuous in three-dimensional Euclidean space R_3 belong to class K, i.e., $p(x) \in K$, if we have $\lim_{|x|\to\infty} \int_{S_a} |p(y)|\, dy = 0$ for a positive. Here $S_a(x)$ is a sphere of radius a with center at the point x;

3) p(x) belongs to class Q_1, i.e., $p(x) \in Q_1$, if we have $\lim_{\rho\to\infty} \rho \int_\rho^\infty |p(r, \omega)|\, dr = 0$. The limit is a uniform one with respect to ω. Here (r, ω) are the spherical coordinates of the point x.

2. With a suitable choice of units, the Dirac operator with a scalar potential in R_3 can be written as (for example, see [2])

$$S\varphi = \sum_{k=1}^{3} \alpha_k p_k \varphi + \alpha_4 \varphi + q(x)\varphi. \tag{1.1}$$

Here, $p_k = -i(\partial/\partial x_k)$ $(k = 1, 2, 3)$, $\varphi(x) = \{\varphi_k(x)\}_{k=1}^{k=4}$ is a four-component function (a bispinor), and α_k $(k = 1, 2, 3, 4)$ are the Dirac matrices satisfying the relation $\alpha_j \alpha_k + \alpha_k \alpha_j = 2\delta_j^k$. For example, they can be chosen as (see [3])

$$\alpha_1 = \begin{pmatrix} 0&0&0&1\\ 0&0&1&0\\ 0&1&0&0\\ 1&0&0&0 \end{pmatrix}, \quad \alpha_2 = \begin{pmatrix} 0&0&0&-i\\ 0&0&i&0\\ 0&-i&0&0\\ i&0&0&0 \end{pmatrix}, \quad \alpha_3 = \begin{pmatrix} 0&0&-1&0\\ 0&0&0&1\\ 1&0&0&0\\ 0&-1&0&0 \end{pmatrix}, \quad \alpha_4 = \begin{pmatrix} 1&0&0&0\\ 0&1&0&0\\ 0&0&-1&0\\ 0&0&0&-1 \end{pmatrix}. \tag{1.2}$$

We will assume that the scalar potential q(x) is bounded.

The operator (1.1) with $q \equiv 0$ will be denoted by S_0; let us also introduce the abbreviation

$$S' = \sum_{k=1}^{3} \alpha_k p_k. \tag{1.3}$$

The spectrum of the self-adjoint operator S_0 fills the whole of the axis with the exception of the interval $(-1, 1)$ (see [2] or [4]). It is also well known that

$$\|S'\varphi\|^2 = \int |\nabla\varphi|^2\, dx = \int \sum_{k=1}^{4} \sum_{j=1}^{3} \left|\frac{\partial \varphi_k}{\partial x_j}\right|^2 dx, \tag{1.4}$$

$$\|S_0\varphi\|^2 = \|S'\varphi\|^2 + \|\varphi\|^2 = \int |\nabla\varphi|^2\, dx + \int |\varphi|^2\, dx. \tag{1.5}$$

3. In the case of spherical symmetry, the Dirac operator after a separation of variables transform as follows (for example, see [2]):

$$F\psi = \rho_1 \frac{\partial \psi}{\partial r} + \rho_2 \psi - \frac{k\rho_3}{r}\psi + q(r)\psi, \quad \psi(0) = 0, \tag{1.6}$$

where $\psi = \{\psi_1(r), \psi_2(r)\}$ is a two-component function, k is one of the numbers ±1, ±2, ±3, . . ., and the matrices ρ_1, ρ_2, and ρ_3 can be chosen as

$$\rho_1 = \begin{pmatrix} 0&1\\ -1&0 \end{pmatrix}, \quad \rho_2 = \begin{pmatrix} -1&0\\ 0&1 \end{pmatrix}, \quad \rho_3 = \begin{pmatrix} 0&1\\ 1&0 \end{pmatrix}.$$

These matrices satisfy the following relations:

$$\left.\begin{aligned} &\rho_i\rho_j = -\rho_j\rho_i \ (i \neq j), \ \rho_2^2 = \rho_3^2 = -\rho_1^2 = 1, \\ &\rho_1^* = -\rho_1, \ \rho_2^* = \rho_2, \ \rho_3^* = \rho_3, \ \rho_2\rho_3 = -\rho_1, \\ &\rho_1\rho_2 = \rho_3, \ \rho_3\rho_1 = \rho_2. \end{aligned}\right\} \tag{1.7}$$

Let us denote the operator F for $q \equiv 0$ by F_0 and let us take $F'\psi = \rho_1 \frac{d\psi}{dr} - k\frac{\rho_3}{r}\psi$.

The spectrum of the operator F_0 is continuous and fills the whole of the axis with the exception of the interval $(-1, 1)$.

With the help of (1.7), we can easily check that

$$\|F_0\psi\|^2 = \int_0^\infty \left(\left| \frac{d\psi}{dr} \right|^2 + |\psi|^2 + \frac{k^2}{r^2}|\psi|^2 \right) dr - k\left(\frac{\rho_2\psi}{r^2}, \psi \right), \tag{1.8}$$

$$\|F_0\psi\|^2 = \|F'\psi\|^2 + \|\psi\|^2. \tag{1.9}$$

4. Let us introduce a Dirac operator with a positive parameter h. In the three-dimensional case it will look as follows:

$$S(h) = hS' + \alpha_4 + q(x).$$

In a similar manner, for the spherically symmetric case we introduce the operator

$$F(h) = hF' + \rho_2 + q(r).$$

We will also need the following Schroedinger operators

$$\left.\begin{aligned} D_1(h, \beta) &= -h^2\Delta + \beta q^2(x) - \frac{2\beta}{\beta+1} q(x), \\ D_2(h, \beta) &= -h^2\Delta + \beta q^2(x) + \frac{2\beta}{\beta+1} q(x), \end{aligned}\right\} \quad (\beta > 0)$$

in $L_2(R_3)$ and

$$\left.\begin{aligned} A_1(\beta)y &= -y'' + q_1(\beta, r)y, \\ A_2(\beta)y &= -y'' + q_2(\beta, r)y, \end{aligned}\right\} \quad y(0) = 0$$

in $L_2(0, \infty)$, where

$$\left.\begin{aligned} q_1(\beta, r) &= \frac{k(k+1)}{r^2} + \beta q^2(r) - \frac{2\beta}{\beta+1} q(r), \\ q_2(\beta, r) &= \frac{k(k-1)}{r^2} + \beta q^2(r) + \frac{2\beta}{\beta+1} q(r). \end{aligned}\right\} \quad (\beta > 0) \tag{1.10}$$

§ 2. The Discrete Spectrum of the Dirac Operator in the Case of Spherical Symmetry

The necessary condition for the total multiplicity of the spectrum of the Dirac operator F(h) to be infinite is provided by the following theorem:

THEOREM 1. *If the total multiplicity of the spectrum of the Dirac operator F(h) with a bounded potential is infinite in the interval (−1, 1) for all positive values of h, then the following condition holds:*

$$\sup_r r\int_r^\infty |q|\,dr = +\infty.$$

Theorem 1 is proved with the help of an upper bound to the total multiplicity of the spectrum of F(h) in the interval (−1, 1) obtained in exactly the same way as the bound contained in Theorem 5.5 of [1]. The presence of the parameter h in the operator F(h) does not affect the derivation of the bound. Next, we make use of Theorem 4.4 of [1] concerning the necessary and sufficient conditions for the unboundedness to the left of zero of the total multiplicity of the spectrum of the Schroedinger operator with a parameter and also the fact that the potential q is bounded.

Let us now establish the necessary condition for the total multiplicity of the spectrum of the Dirac operator F in the interval $(-1, 1)$ to be infinite. For this purpose, let us denote the total multiplicities of the negative spectra of the operators $A_1(\beta)$ and $A_2(\beta)$ by N_1 and N_2, respectively. N_1 coincides with the maximum dimensionality of the linear sets of finite* function ψ_1 satisfying the inequality

$$\int_0^\infty \left(\left|\frac{d\psi_1}{dr}\right|^2 + \frac{k(k+1)}{r^2}|\psi_1|^2 + \beta q^2|\psi_1|^2 - \frac{2\beta}{\beta+1}q|\psi_1|^2\right)dr < 0. \tag{2.1}$$

N_2 coincides with the maximum dimensionality of the linear sets of finite functions ψ_2 satisfying the inequality

$$\int_0^\infty \left(\left|\frac{d\psi_2}{dr}\right|^2 + \frac{k(k-1)}{r^2}|\psi_2|^2 + \beta q^2|\psi_2|^2 + \frac{2\beta}{\beta+1}q|\psi_2|^2\right)dr < 0. \tag{2.2}$$

It follows from this that the maximum dimensionality of linear sets of finite functions of the form (ψ_1, ψ_2) satisfying the inequality

$$\int_0^\infty \left(\left|\frac{d\psi_1}{dr}\right|^2 + \frac{k(k+1)}{r^2}|\psi_1|^2 + \beta q^2|\psi_1|^2 - \frac{2\beta}{\beta+1}q|\psi_1|^2 + \left|\frac{d\psi_2}{dr}\right|^2 + \right.$$
$$\left. + \frac{k(k-1)}{r^2}|\psi_2|^2 + \beta q^2|\psi_2|^2 + \frac{2\beta}{\beta+1}q|\psi_2|^2\right)dr < 0 \tag{2.3}$$

is not less than $N_1 + N_2$. Inequality (2.3) can be rewritten as

$$\left\|\frac{d\psi}{dr}\right\|^2 + k^2\left\|\frac{\psi}{r}\right\|^2 - k\left(\rho_2\frac{\psi}{r}, \frac{\psi}{r}\right) + \frac{2\beta}{\beta+1}(\rho_2 q\psi, \psi) + \beta\|q\psi\|^2 < 0,$$

where $\psi = (\psi_1, \psi_2)$. With the help of (1.8) and (1.9), we can rewrite the last inequality as

$$\|F'\psi\|^2 < -\frac{2\beta}{\beta+1}(\rho_2\psi, q\psi) - \beta\|q\psi\|^2. \tag{2.4}$$

Taking account of the elementary inequality

$$\|F'\psi\|^2 + \beta\|q\psi\|^2 \geq \frac{\beta}{\beta+1}\|F'\psi + q\psi\|^2, \quad \beta > 0, \tag{2.5}$$

we find from (2.4) that

$$\|F'\psi + q\psi\|^2 + 2(\rho_2\psi, q\psi) < 0$$

* We have in mind finiteness near zero and at infinity.

or

$$\|F'\psi\|^2 + 2\,\mathrm{Re}\,(F'\psi,\ q\psi) + \|q\psi\|^2 + 2(p_2\psi,\ q\psi) < 0,$$

which, together with (1.9), leads to

$$\|F\psi\|^2 < \|\psi\|^2. \tag{2.6}$$

Thus, we come to the conclusion that the total multiplicity of the spectrum of F in the interval $(-1, 1)$ is not less than $N_1 + N_2$. Consequently, we have

THEOREM 2. The total multiplicity of the spectrum of the operator F in the interval $(-1, 1)$ is not less than the sum of the total multiplicities of the negative spectra of the operators $A_1(\beta)$ and $A_2(\beta)$ for any positive β.

With the help of Theorem 2, it is not difficult to establish the sufficient condition for the unboundedness of the total multiplicity of the spectrum of the operator F in the interval $(-1, 1)$. For this purpose, let us assume that we can find an r_0 such that $q(r)$ preserves its sign and $|q| \leq \varepsilon < 2$ when $r > r_0$. Let us also assume that

$$|q|\,r^2 \to \infty \quad \text{or} \quad r \to \infty. \tag{2.7}$$

Then, it is obvious that

$$\sup_{r>r_0} r \int_r^\infty |q|\,dr = +\infty. \tag{2.8}$$

Making use of the condition that $|q| \leq \varepsilon < 2$ for $r > r_0$, we can easily establish that one of the potentials $q_1(r, \beta)$ or $q_2(r, \beta)$, introduced by formula (1.10), is negative for $r > r_0$ provided that $0 < \beta < 2\varepsilon^{-1} - 1$. Making use of (2.8), we conclude that for such values of β we have either

$$\sup_{r>r_0} r \int_r^\infty (-q_1(\beta,\ r))\,dr = +\infty \tag{2.9}$$

or

$$\sup_{r>r_0} r \int_r^\infty (-q_2(\beta,\ r))\,dr = +\infty. \tag{2.10}$$

On the basis of Theorem 4.4 of [1], we conclude that the total negative spectrum of the operator $A_1(\beta)$ or $A_2(\beta)$ is unbounded when $0 < \beta < 2\varepsilon^{-1} - 1$. This result, together with Theorem 2, shows that the total spectrum of the operator F in the interval $(-1, 1)$ is also unbounded. Consequently, we have

THEOREM 3. If there exists a number r_0 such that $q(r)$ preserves its sign and $|q| \leq \varepsilon < r_0$ and if $|q|\, r^2 \to \infty$ as $r \to \infty$, then the total multiplicity of the spectrum of the operator F in the interval $(-1, 1)$ is unbounded.

From Theorem 3 and Theorem 5.1 of [1] concerning the condition for the spectrum of the Dirac operator in $(-1, 1)$ to be discrete, it now follows that if we add the condition $q \in K$ to the conditions of Theorem 3, then we can assert that the total multiplicity of the spectrum of the operator F in $(-1, 1)$ is unbounded and that the spectrum in the interval $(-1, 1)$ is a purely discrete spectrum. In particular, this is so if we have $q = Cr^{\alpha-2}$, where $0 < \alpha < 2$ and C is an arbitrary real number. On the basis of the results obtained in [1] (see Theorems 5.2 and 5.4), we can conclude that the accumulation point of the eigenvalues will be the point $\lambda = -1$ when $C > 0$ and the point $\lambda = +1$ when $C < 0$.

§ 3. The Discrete Spectrum of the Dirac Operator in the Three-Dimensional Case

In this section we will study the discrete spectrum of the operator S(h). Let us first of all derive the upper bound to the total multiplicity of the spectrum of the operator S in (−1, 1). Our result will be an improvement of Theorem 5.2 of [1].

The total multiplicity of the spectrum of the Dirac operator S in (−1, 1) coincides with the maximum dimensionality of the linear sets of finite functions satisfying the inequality

$$\|S\psi\| < \|\psi\|. \tag{3.1}$$

Following the derivation given in [1] and making use of the explicit form of the matrices α_4, we easily find from inequality (3.1) that

$$\int\left[|\nabla\psi'|^2 - \left(\frac{\beta+1}{\beta}q^2 + 2(\beta+1)q\right)|\psi'|^2\right]dx + \int\left[|\nabla\psi''|^2 - \left(\frac{\beta+1}{\beta}q^2 - 2(\beta+1)q\right)|\psi''|^2\right]dx < 0, \quad \beta > 0, \tag{3.2}$$

where $\psi' = (\psi_3, \psi_4)$ and $\psi'' = (\psi_1, \psi_2)$. This leads to

THEOREM 4. *The total multiplicity of the spectrum of the Dirac operator S in (−1, 1) does not exceed twice the sum of the total multiplicities of the negative spectra of the Schroedinger operators with potentials*

$$-(\beta+1)\beta^{-1}q^2 - 2(\beta+1)q \quad \text{and} \quad -(\beta+1)\beta^{-1}q^2 + 2(\beta+1)q \quad (\beta > 0).$$

Let us establish the necessary condition for the boundedness of the spectrum of the Dirac operator S(h) in (−1, 1). For this, we first of all prove two lemmas giving lower bounds to the total multiplicity of the operator S(h) in (−1, 1).

The total multiplicity of the negative spectrum of the operator $D_1(h, \beta)$ coincides with the maximum dimensionality of the linear sets of finite functions satisfying the inequality

$$h^2\int|\nabla\varphi|^2dx < -\beta\int q(q - 2(\beta+1)^{-1})|\varphi|^2dx. \tag{3.3}$$

For the linear manifold of functions $\psi = (0, 0, \varphi, \varphi)$, we have

$$h^2\int|\nabla\psi|^2dx < -\beta\int q(q - 2(\beta+1)^{-1})|\psi|^2dx$$

or, equivalently,

$$(\beta+1)\|hS'\psi\|^2 < -\beta(\beta+1)\|q\psi\|^2 - 2\beta\,\mathrm{Re}(\alpha_4\psi, q\psi),$$

where S' and α_4 are defined by (1.2) and (1.3). Making use of the inequality analogous to (2.5), but written down for S', we obtain

$$\|hS'\psi + q\psi\|^2 < -\mathrm{Re}(\alpha_4\psi, q\psi),$$

or

$$\|hS'\psi\|^2+2\,\mathrm{Re}\,(hS'\psi,\ q\psi)+\|q\psi\|^2+2\,\mathrm{Re}\,(\alpha_4\psi,\ q\psi)<0.$$

Next, in view of the definition of S_0, we have

$$\|S_0(h)\psi\|^2-\|\psi\|^2+2\,\mathrm{Re}\,(S_0(h)\psi,\ q\psi)+\|q\psi\|^2<0,$$

which yields

$$\|S(h)\psi\|^2<\|\psi\|^2.$$

It should be noted that the dimensionality of the linear manifold of functions ψ coincides with that of the manifold of functions φ. Consequently, we have

LEMMA 1. The total multiplicity of the spectrum of the Dirac operator $S(h)$ in $(-1, 1)$ is not less than the total multiplicity of the negative spectrum of the Schroedinger operator $D_1(h, \beta)$.

If the functions ψ are replaced by $\psi = (\varphi, \varphi, 0, 0)$, then we can analogously prove

LEMMA 2. The total multiplicity of the spectrum of the Dirac operator $S(h)$ in $(-1, 1)$ is not less than the total multiplicity of the negative spectrum of the Schroedinger operator $D_2(h, \beta)$.

Let us now assume that for all positive h the Dirac operator S(h) in $(-1, 1)$ has a bounded spectrum. Then, on the basis of Lemmas 1 and 2, the negative spectrum of the operators $D_1(h, \beta)$ and $D_2(h, \beta)$ is bounded. If, in addition, the potential q does not change sign and $|q| \leq \varepsilon < 2$, then the following operator has a bounded negative spectrum:

$$-h^2\Delta-\beta|q|\left(\varepsilon-\frac{2}{\beta+1}\right). \tag{3.4}$$

When $\beta < 2\varepsilon^{-1} - 1$, then potential of operator (3.4) is positive. Comparing this result with Theorem 5.3 of [1] and making use of Theorem 4.5 of [1], we arrive at the following theorem.

THEOREM 5. If for sufficiently large $|x|$ the potential q is bounded in absolute magnitude, then under the condition $|q| \in Q_1$ the operator S has a bounded spectrum in $(-1, 1)$. If for all positive h the operator S(h) in $(-1, 1)$ has a bounded spectrum, the potential q does not change sign, is almost radial, and $|q| \leq \varepsilon < 2$, then we have $|q| \in Q_1$.

On the basis of Lemmas 1 and 2 and Theorem 4.7 of [1] on the necessary and sufficient conditions for the negative spectrum of the Schroedinger operator to be unbounded, we can state

THEOREM 6. If the potential q does not change sign, $|q| \leq \varepsilon < 2$ for sufficiently large $|x|$, and

$$\lim_{b\to\infty} b\int_{|x|>b}|q|\,|x|^{-2}\,dx=\infty,$$

then the spectrum of the Dirac operator S in $(-1, 1)$ is unbounded.

Let us note in conclusion that making use of the quantitative bounds on the total multiplicity of the negative spectrum of the Schroedinger operator (see [1]), we can obtain analogous bounds on the total multiplicity of the Dirac operator in (−1, 1).

§ 4. The Discrete Spectrum of the Pauli Operator

In this paragraph we will obtain results concerning the negative part of the spectrum of the Pauli operator. This operator can be written as (for example, see [4, 5])

$$P\psi = P_0\psi + Q(x)\psi, \tag{4.1}$$

where

$$P_0\psi = -\Delta\psi - 2i\sum_{j=1}^{3} A_j(x)\frac{\partial\psi}{\partial x_j} + |A|^2\psi,$$

$$Q(x) = \begin{pmatrix} H_3+q & H_1+iH_2 \\ H_1-iH_2 & -H_3-q \end{pmatrix}, \quad \psi = \begin{pmatrix}\psi_1\\ \psi_2\end{pmatrix}, \tag{4.2}$$

$$|A|^2 = \sum_{j=1}^{3} |A_j(x)|^2.$$

Here, $\mathbf{A} = (A_1, A_2, A_3)$ is the vector potential and $\mathbf{H} = (H_1, H_2, H_3)$ is the vector of the magnetic field intensity. We have the following relations: $\mathbf{H} = \operatorname{rot} \mathbf{A}$ and $\operatorname{div} \mathbf{A} = 0$.

Let $\nu_1(x)$ and $\nu_2(x)$ denote the largest and smallest eigenvalues of the matrix $Q(x)$,

$$\nu_1(x) = q(x) + H(x), \quad \nu_2(x) = q(x) - H(x),$$

where

$$H(x) = \sqrt{H_1^2 + H_2^2 + H_3^2}.$$

The total multiplicity of the negative spectrum of the operator P is equal to the maximum dimensionality of the linear manifolds of finite functions satisfying the inequality

$$(P\psi, \psi) < 0. \tag{4.3}$$

If inequality (4.3) is satisfied, then we have

$$(-\Delta\psi, \psi) - 2\left(\sum_{j=1}^{3} |A_j(x)|\cdot\left|\frac{\partial\psi}{\partial x_j}\right|, |\psi|\right) + (A^2\psi, \psi) + (Q\psi, \psi) < 0.$$

Here, $|\psi| = (|\psi_1|, |\psi_2|)$, and the $\left|\frac{\partial\psi}{\partial x_j}\right| = \left(\left|\frac{\partial\psi_1}{\partial x_j}\right|, \left|\frac{\partial\psi_2}{\partial x_j}\right|\right)$ are vectors. Consequently, we have

$$\sum_{j=1}^{2}\left[\int (|\nabla\psi_j|^2 + |A|^2|\psi_j|^2)\,dx - 2\sqrt{\int |A|^2|\psi_j|^2 dx}\ \sqrt{\int |\nabla\psi_j|^2 dx} + \int (q-H)|\psi_j|^2 dx\right] < 0. \tag{4.4}$$

Making use of Cauchy's inequality, we obtain from (4.4)

$$\sum_{j=1}^{2}\int\left[|\nabla\psi_j|^2 - \left(\varepsilon^{-1}|A|^2 + \frac{H-q}{1-\varepsilon}\right)|\psi_j|^2\right]dx < 0 \quad (0<\varepsilon<1). \tag{4.5}$$

Consequently, we have

LEMMA 3. The total multiplicity of the negative spectrum of the Pauli operator does not exceed twice the total multiplicity of the negative spectrum of the Schroedinger operator with potential of the form $\varepsilon^{-1}|A|^2 - (H - q)(1 - \varepsilon)^{-1}$ for any ε in the interval (0, 1).

Let us now consider the Schroedinger operator

$$-\Delta\Phi + (\varepsilon^{-1}|A|^2 + (H+q)(1+\varepsilon)^{-1})\Phi, \quad \varepsilon > 0.$$

The total multiplicity of the negative spectrum of this operator is equal to the maximum dimensionality of the linear sets of finite functions satisfying the inequality

$$\int [|\nabla\Phi|^2 + (\varepsilon^{-1}|A|^2 + (H+q)(1+\varepsilon)^{-1})|\Phi|^2]\, dx < 0$$

or

$$\int \left[|\nabla\Phi|^2 + \varepsilon|\nabla\Phi|^2 + |A|^2|\Phi|^2 + \frac{1}{\varepsilon}|A|^2|\Phi|^2 + (H+q)|\Phi|^2\right] dx < 0.$$

Using the Cauchy inequality, we now find that

$$\int [|\nabla\Phi|^2 + |A|^2|\Phi|^2]\, dx + 2\sqrt{\int |A|^2|\Phi|^2 dx}\sqrt{\int |\nabla\Phi|^2 dx} + \int (q+H)|\Phi|^2 dx < 0. \tag{4.6}$$

It follows from inequality (4.6) that $(P\psi, \psi) < 0$ if we assume that $\psi = (\Phi, 0)$ or $\psi = (0, \Phi)$. This leads to

LEMMA 4. Twice the total multiplicity of the negative spectrum of the Schroedinger operator with potential $\varepsilon^{-1}|A|^2 + (H + q)(1 + \varepsilon)^{-1}$ does not exceed the total multiplicity of the negative spectrum of the Pauli operator P. Here, ε is an arbitrary positive number.

If we impose additional restrictions on $|A|$, q, and H, we can derive from Lemmas 3 and 4 simple conditions for the negative spectrum of the Pauli operator to be unbounded. Let us give some of them.

THEOREM 7. Let us suppose that for sufficiently large $|x|$ and a positive C we have $|A|^2 \leq C|q|$ and $H \leq C|q|$. Then, under the condition $q_- \in Q_1$, the Pauli operator

$$P(h)\psi = -h^2\Delta\psi - 2ih\sum_{j=1}^{3} A_j(x)\frac{\partial\psi}{\partial x_j} + |A|^2\psi + Q\psi$$

has a bounded negative spectrum. If for all positive h, the operator P(h) with an almost radial potential $|q(x)|$ has a bounded negative spectrum, $q(x) < 0$, and for a positive C we have $|A|^2 \leq C|q|$ and $H \leq C|q|$, then $|q(x)| \in Q_1$. Here, $q_-(x) = \min\{q(x); 0\}$.

THEOREM 8. Let us suppose that for sufficiently large $|x|$ the potential q is negative and that for some positive C we have $|A|^2 \leq C|q|$ and $H \leq C|q|$. Then, under the condition

$$\varlimsup_{b\to\infty} b \int_{|x|\geq b} q_-(x)\cdot|x|^{-2} dx < +\infty$$

the negative spectrum of the operator P(h) is unbounded. If for all real h the operator P(h) has an unbounded negative spectrum and the potential q(x) is radial, then condition (4.8) is satisfied.

LITERATURE CITED

1. M. Sh. Birman, "The spectra of singular boundary problems," Matem. Sbornik, Vol. 55 (97) (1961).
2. V. A. Fok, The Origins of Quantum Mechanics, Izd. LGU (1932).
3. H. A. Bethe and E. E. Salpeter, Quantum Mechanics of One and Two Electron Atoms Academic Press, New York (1957).
4. I. M. Glazman, Direct Methods for the Spectral Analysis of Singular Differential Operators, Izd. Nauka (1963).
5. L. D. Landau and E. M. Lifshits, Quantum Mechanics (Nonrelativistic Theory, Fizmatgiz (1963).
6. B. Ya. Skachek, "A remark concerning the spectrum of the Dirac operator," Dokl. Akad. Nauk Ukrain.SSR, Vol. 65 (10) (1965).

THE NONSELF-ADJOINT SCHROEDINGER OPERATOR. III

B. S. Pavlov

The present article contains a detailed derivation of some results previously summarized in a note by the author [1]. It is also a continuation of the article published in the second volume of the present series [2]. An example of a nonself-adjoint Schroedinger operator with a rapidly decreasing potential and an infinite number of eigenvalues was given in [2]. Here, we will show that the spectrum of the Schroedinger operator can have a very complicated structure. Namely, there exist operators of this type whose eigenvalues possess a continuum of accumulation points. These results have been formulated as Theorems I, II, and III.

§ 1. Auxiliary Information

Everywhere in the following we consider the operator l generated in $L_2(0, \infty)$ by the differential expression

$$ly = -y'' + q(x)y \tag{1.1}$$

and the boundary condition

$$y' - hy(0) = 0. \tag{1.2}$$

Here, h is a complex number and q(x) is a bounded continuous real function. We will say that it belongs to class S_n, n = 0, 1, 2, . . ., if $\int_0^\infty |q(x)| x^{n+1} dx < \infty$; moreover, we have $q(x) \in S_\infty$, if q(x) belongs to S_n for any n.

Let us consider an equation containing a complex parameter λ

$$-y'' + q(x)y = \lambda y. \tag{1.3}$$

Let $\varphi_h(x, \lambda)$ and $\psi_h(x, \lambda)$ denote the solutions of Eq. (1.3) satisfying the initial conditions

$$\left.\begin{aligned} \varphi_h(0, \lambda) &= (1+h^2)^{-\frac{1}{2}}, \quad \varphi_h'(0, \lambda) = h(1+h^2)^{-\frac{1}{2}}, \\ \psi_h(0, \lambda) &= -h(1+h^2)^{-\frac{1}{2}}, \quad \psi_h'(0, \lambda) = (1+h^2)^{-\frac{1}{2}}. \end{aligned}\right\} \tag{1.4}$$

The general solution of Eq. (1.3) is of the form

$$\psi_h(x, \lambda) + m\varphi_h(x, \lambda).$$

Let us determine the value of the number $m = m_h(\lambda)$ from the condition

$$\psi_h(x, \lambda) + m_h(\lambda)\varphi_h(x, \lambda) \equiv \chi(x, \lambda) \in L_2(0, \infty).$$

This can be done in a unique manner for any nonreal value of λ. The function $m_h(\lambda)$ has been introduced by Weyl [3]. We call it the Weyl function of the operator l_h. When h is a real quantity, the Weyl function is regular in each of the half-planes $\operatorname{Im}\lambda > 0$, $\operatorname{Im}\lambda < 0$ and satisfies the condition

$$\operatorname{Im} m_h(\lambda) \operatorname{Im} \lambda < 0.$$

In the self-adjoint case ($\operatorname{Im} h = 0$), the Weyl function is simply related to the spectral function of the operator l_h (see [4]). The analogous relation for the nonself-adjoint case has been established in [5]. The Weyl functions corresponding to different boundary conditions are related by

$$\frac{-h_1 + m_{h_1}(\lambda)}{h_1 m_{h_1}(\lambda) + 1} = \frac{-h_2 + m_{h_2}(\lambda)}{h_2 m_{h_2}(\lambda) + 1}. \tag{1.5}$$

If we know the Weyl function $m_h(\lambda)$, we can give a complete description of the spectrum of l_h. The poles of the Weyl function coincide with eigenvalues, while the other singularities define the limiting spectrum of l_h.

In the proofs of the assertions of the present article, we will make use of Lemma 2.1 of [2]. In the following it will be called the Principal Lemma.

THE PRINCIPAL LEMMA. Let $m(k)$ be a function that is regular in some neighborhoods of the points $k = 0$ and $k = \infty$ and in the half-plane $\operatorname{Im} k > 0$; let $m(k)$ possess the following properties:

(1) $m(k)$ is continuous right up to the real axis together with all of its derivatives of order equal to or less than n, $n \geq 3$,

(2) $m(0) \neq 0$, $m'(0) \neq 0$,

(3) $km(k) \to -i$ as $|k| \to \infty$,

(4) $m(k)$ is real on the positive imaginary axis, while the following condition holds on the real axis:

$$\operatorname{Im} m(-k) = -\operatorname{Im} m(k) > 0 \quad \text{for } k > 0.$$

Then, there exists an infinitely differentiable real function $q(x)$ belonging to S_{n-3} and a real number a_0 such that

$$-\left[m(\sqrt{\lambda})\right]^{-1} + a_0$$

is a Weyl function of the differential operator l_∞ with a potential $q(x)$ and the boundary condition $y(0) = 0$ ($h = \infty$).

§2. The Operator with Potential $q(x) \in S_\infty$

The connection between the rate of decrease of the potential q(x) and the structure of the accumulation-point set of the eigenvalues of the operator l_h has been established in [1] and [6]. This connection made possible the accumulation of eigenvalues on a set with the power of the continuum. On the other hand, an example of an operator whose eigenvalues have only one accumulation point has been given in [2]. This left the following question open: Can the eigenvalues of the operator l_h with potential $q(x) \in S_\infty$ really accumulate on a point set with the structure described in [1, 6]?

A partial* answer to this question is provided by the theorem proved in this section. Let us formulate it.

Let us agree to say that two bounded sets on the real axis are similar if the bounded intervals containing them can be mapped onto one another in a continuously differentiable manner such that one set transforms into the other.

THEOREM 1. Let **E** be an arbitrary bounded closed point set of the real axis of measure zero satisfying the condition

$$\sum |l_\nu|^{1-\gamma} < \infty, \quad 1 > \gamma > 0, \tag{2.6}$$

where $|l_\nu|$ is the length of the interval l_ν of contiguity with the set **E** and the summation extends over all bounded intervals l_ν.

Then, there exists a differential operator of the form of l_h with an infinitely differentiable real potential $q(x) \in S_\infty$ such that for some complex boundary condition the accumulation-point set for the eigenvalues of l_h is similar to **E**.

PROOF. As in the case of Theorem 1 of [2], the proof consists in the construction of a function m(k), satisfying the conditions of the Principal Lemma.

We can assume without loss of generality that the set **E** is situated on the interval $\pi/2 \le \vartheta \le 3\pi/2$, where $\pi/2, 3\pi/2 \in$ **E**. Let us denote by a_ν, b_ν ($a_\nu < b_\nu$), $\nu \ge 1$, the end points of the intervals of contiguity with the set **E**. Assuming that $a_0 = -\pi/2$, $b_0 = \pi/2$, and that ϑ is a polar angle, we can transform the set **E** into a unit circle.

Let us introduce the function† $h(\vartheta)$, where $\vartheta \in \left(-\frac{\pi}{2}, \frac{3\pi}{2}\right]$ defined by

$$h(\vartheta) = \begin{cases} (b_\nu - \vartheta)^{-\gamma} + (\vartheta - a_\nu)^{-\gamma} \equiv h_\nu(\vartheta), & \vartheta \in (a_\nu, b_\nu), \\ +\infty & \text{at all other points} \end{cases} \tag{2.7}$$

$h(\vartheta)$ is analytic in the intervals (a_ν, b_ν) and is summable on the unit circle. The latter property

---

* In the example constructed by us we have $q(x) \in S_\infty$. It appears that the inequality $|q(x)| \le c\exp\left[-dx^{\frac{\gamma'}{1+\gamma'}}\right]$ is also valid for some $\gamma' > 0, \gamma' < \gamma$. However, we do not possess a proof of this assertion.

† Functions of this type have been used by many authors for a variety of reasons (for example, see [7]). We need to study in greater detail the properties of functions regular in the unit circle and associated with $h(\vartheta)$.

follows from

$$\int_{-\frac{\pi}{2}}^{\frac{3\pi}{2}} |h(\vartheta)|\, d\vartheta = \sum_{\nu=0}^{\infty} \int_{a_\nu}^{b_\nu} h_\nu(\vartheta)\, d\vartheta = \frac{2}{1-\gamma}\left[\sum_{\nu=1}^{\infty} |l_\nu|^{1-\gamma} + \pi^{1-\gamma}\right] < \infty.$$

The Poisson integral

$$u(r, \varphi) = \frac{1}{2\pi} \int_{-\frac{\pi}{2}}^{\frac{3\pi}{2}} \frac{(1-r^2)\, h(\vartheta)}{1+r^2-2r\cos(\varphi-\vartheta)}\, d\vartheta, \quad r<1, \ -\frac{\pi}{2} < \varphi \leqslant \frac{3\pi}{2}$$

yields a function that is harmonic in the unit circle. In view of the maximum principle, the following inequality holds at all points interior to the unit circle:

$$u(r, \varphi) > \inf h(\vartheta) = h(0) = 2^{\gamma-1}\pi^{\gamma} > 0. \tag{2.8}$$

Let us find the lower bound to $u(r, \varphi)$ in the neighborhood of **E**.

Let ϑ_0 be an element of **E** and let D_ε be a domain defined by the inequalities $|\varphi - \vartheta_0| < \varepsilon_1$, $|r-1| < \varepsilon_2$, $r < 1$. We will assume that the numbers ε_1 and ε_2 are so small that the inequalities $(1-\varepsilon_2)(1-\cos\varepsilon_1) > 4^{-1}\varepsilon_1^2$, $1 - 4\varepsilon_2^{1/2} > 2^{-1}$, and $\varepsilon_1^{-1}\sin\varepsilon_1^{-1}$ are satisfied. Then, for $(r, \varphi) \in D_\varepsilon$, we have

$$0 < \frac{1}{2\pi} \int_{|\vartheta-\vartheta_0| > 2\varepsilon_1} \frac{1-r^2}{1+r^2-2r\cos(\varphi-\vartheta)}\, d\vartheta =$$

$$= \frac{1}{2\pi} \int_{|\vartheta-\vartheta_0| > 2\varepsilon_1} \frac{(1+r)(1-r)}{(1-r)^2+2r\,[1-\cos(\varphi-\vartheta)]}\, d\vartheta < \frac{2\varepsilon_2}{2(1-\varepsilon_2)[1-\cos\varepsilon_1]} < 4\frac{\varepsilon_2}{\varepsilon_1^2}.$$

Taking into account that

$$\frac{1}{2\pi} \int_{-\frac{\pi}{2}}^{\frac{3\pi}{2}} \frac{1-r^2}{1+r^2-2r\cos(\varphi-\vartheta)}\, d\vartheta = 1, \quad r < 1,$$

we find from $(r, \varphi) \in D_\varepsilon$ that

$$\frac{1}{2\pi} \int_{|\vartheta-\vartheta_0| \leqslant 2\varepsilon_1} \frac{1-r^2}{1+r^2-2r\cos(\varphi-\vartheta)}\, d\vartheta \geqslant 1 - 4\frac{\varepsilon_2}{\varepsilon_1^2}.$$

Since $h(\vartheta) \geq |\vartheta - \vartheta_0|^{-\gamma}$ for $\vartheta_0 \in E$, $\vartheta \,\bar{\in}\, E$, the function $u(r, \varphi)$ has the following lower bound in D_ε:

$$u(r, \varphi) > \frac{1}{2\pi} \int_{|\vartheta-\vartheta_0| \leqslant 2\varepsilon} \frac{(1-r^2)\, h(\vartheta)}{1+r^2-2r\cos(\varphi-\vartheta)}\, d\vartheta \geqslant \frac{1}{2\pi} \int_{|\vartheta-\vartheta_0| < 2\varepsilon} \frac{(1-r^2)\, d\vartheta}{1+r^2-2r\cos(\varphi-\vartheta)} \cdot \frac{1}{2^{\gamma}\varepsilon_1^{\gamma}} \geqslant \frac{1}{2^{\gamma}\varepsilon_1^{\gamma}}\left[1 - 4\frac{\varepsilon_2}{\varepsilon_1^2}\right].$$

Let dist $[(r, \varphi), \mathbf{E}]$ denote the distance of a point with coordinates (r, φ) on the unit circle from the set $\mathbf{E}$. Choosing $\varepsilon_2 = \text{dist}\,[(r, \varphi), \mathbf{E}]$ and $\varepsilon_1 = \varepsilon_2^{1/3}$, we find that in a small neighborhood of $\mathbf{E}$ we have

$$u(r, \varphi) > 2^{1-\gamma}\{\text{dist}\,[(r, \varphi), E]\}^{-\frac{\gamma}{3}}.$$

In view of (2.8), we can assume that for some $K > 0$ the following bound is valid everywhere within the unit circle*

$$u(r, \varphi) > K\{\text{dist}\,[(r, \varphi), E]\}^{-\frac{\gamma}{3}}. \tag{2.9}$$

Let us now introduce a function that is analytic in the unit circle, the real part of this function being the same as $u(r, \varphi)$,

$$f(z) = \frac{1}{2\pi}\int\limits_{-\frac{\pi}{2}}^{\frac{3\pi}{2}} \frac{e^{i\vartheta} + z}{e^{i\vartheta} - z}\, h(\vartheta)\, d\vartheta, \quad |z| < 1. \tag{2.10}$$

This function will be used below for the function m(k). Let us first of all obtain the bounds on the derivatives of $f(z)$. With the help of the definition of $h(\vartheta)$ [Eq. (2.7)], we can rewrite (2.10) in the form of a series

$$f(z) = \sum_{\nu=0}^{\infty} f_\nu(z), \tag{2.11}$$

where

$$f_\nu(z) = \frac{1}{2\pi i}\int\limits_{L_\nu} \frac{x + z}{x - z}\, h_\nu\left(\frac{1}{i}\ln x\right)\frac{dx}{x}. \tag{2.12}$$

Here, L_ν is an arc of the unit circle, $x = e^{i\vartheta}$, $\vartheta \in l_\nu$. The branch of $\ln x$ to be selected is that given by the condition

$$\ln e^{ia_0} = ia_0.$$

In view of

$$|f_\nu(z)| \leq \frac{1 + |z|}{1 - |z|}\,\frac{1}{\pi(1-\gamma)}(b_\nu - a_\nu)^{1-\gamma}$$

and condition (2.6), the series in (2.11) converges absolutely and uniformly in each circle $|z| \leq 1 - \delta$, where $\delta > 0$.

Let us consider the function $f_\nu(z)$ in greater detail. Let us first of all note that the function

$$\tilde{h}_\nu(x) = x^{-1} h_\nu\left(\frac{1}{i}\ln x\right)$$

becomes singly valued and regular in the domain $x > 1$, $a_\nu < \arg x < b_\nu$ when its branch is chosen from the condition

$$\tilde{h}_\nu\left(\exp i\,\frac{b_\nu + a_\nu}{2}\right) = 2^{\gamma-1}|b_\nu - a_\nu|^{-\gamma}\exp\left(-i\,\frac{b_\nu - a_\nu}{2}\right)$$

and after cuts are made along the rays $\arg x = a_\nu$, $\arg x = b_\nu$.

*A minor modification of the argument allows us to replace the exponent in (2.9) by $-\gamma$.

We will show that $\tilde{h}_\nu(x)$ is summable on the segments Γ_ν^1, Γ_ν^3 of the rays $\arg x = a_\nu$, $\arg x = b_\nu$, $1 < |x| < \exp|l_\nu|$, as well as on the arc Γ_ν^3 of the circle $|x| = \exp|l_\nu|$, $a_\nu < \arg x < b_\nu$. Let us evaluate the integral along Γ_ν^1,

$$\int_{\Gamma_\nu^1} |\tilde{h}_\nu(x)|\, d|x| = \int_1^{\exp|l_\nu|} |(\ln\rho)^{-\gamma} + (\ln\rho - i|l_\nu|)^{-\gamma}| \frac{d\rho}{\rho} \leqslant$$

$$\leqslant \int_1^{\exp|l_\nu|} (\ln\rho)^{-\gamma}\frac{d\rho}{\rho} + \int_1^{\exp|l_\nu|} (|l_\nu| - \ln\rho)^{-\gamma}\frac{d\rho}{\rho} = 2(1-\gamma)^{-1}|l_\nu|^{1-\gamma}. \tag{2.13}$$

The integral along Γ_ν^3 has the same value. Further, we have

$$\int_{\Gamma_\nu^2} |\tilde{h}_\nu(x)|\,|x|\, d\arg x = \int_{a_\nu}^{b_\nu} \left|\left[\frac{1}{i}|l_\nu| + \vartheta - a_\nu\right]^{-\gamma} + \left[b_\nu - \vartheta - \frac{1}{i}|l_\nu|\right]^{-\gamma}\right| d\vartheta \leqslant$$

$$\leqslant \int_{a_\nu}^{b_\nu} (\vartheta - a_\nu)^{-\gamma}\, d\vartheta + \int_{a_\nu}^{b_\nu} (b_\nu - \vartheta)^{-\gamma}\, d\vartheta = 2(1-\gamma)^{-1}|l_\nu|^{1-\gamma}. \tag{2.14}$$

In view of what has been said above, the integral in (2.12) can be replaced by an integration along the contour $\Gamma_\nu = \Gamma_\nu^1 + \Gamma_\nu^2 + \Gamma_\nu^3$,

$$f_\nu(z) = \frac{1}{2\pi i}\int_{\Gamma_\nu} \frac{x+z}{x-z}\tilde{h}_\nu(x)\, dx, \quad |z| < 1.$$

In view of (2.13) and (2.14), we have

$$\int_{\Gamma_\nu} |\tilde{h}_\nu(x)|\, d\Gamma_\nu \leqslant 6(1-\gamma)^{-1}|l_\nu|^{1-\gamma}. \tag{2.15}$$

This directly yields

$$|f_\nu(z)| \leqslant \frac{1}{2\pi}\left[\frac{2(1+e^\pi)}{\operatorname{dist}(z, \Gamma_\nu)} + 1\right]\frac{6|l_\nu|^{1-\gamma}}{1-\gamma},$$

$$|f_\nu^{(r)}(z)| \leqslant \frac{6|l_\nu|^{1-\gamma}(1+e^\pi)\, r!}{\pi(1-\gamma)\{\operatorname{dist}(z, \Gamma_\nu)\}^{r+1}}, \quad r > 0. \tag{2.16}$$

It is now an easy matter to obtain the bounds on the derivatives of $f(z)$. First of all, let us note that equality (2.11) gives us the continuation of the function $f(z)$ into the domain bounded by the totality of the contours $\Gamma_0, \Gamma_1, \Gamma_2, \ldots$. The series (2.11) converges absolutely and uniformly within this domain and, in view of (2.16), it can be differentiated term by term any number of times. In view of the inequality $1 + |l_\nu| < \exp|l_\nu|$, the point on the contour Γ_ν lying closest to any point z with $|z| < 1$ is one of the points a_ν or b_ν and, consequently, we have

$$\inf_\nu [\operatorname{dist}(z, \Gamma_\nu)] = \operatorname{dist}(z, E), \quad |z| \leqslant 1. \tag{2.17}$$

For $|z| < 1$, this together with (2.16) yields

$$|f_{(z)}^{(r)}| \leqslant \frac{r!}{[\operatorname{dist}(z, E)]^{r+1}} \frac{6(1+e^\pi)\sum_{\nu=0}^{\infty}|l_\nu|^{1-\gamma}}{\pi(1-\gamma)}, \quad r \geqslant 1. \tag{2.18}$$

Let us now construct the Blaschke function whose zeros accumulate to the points of **E**. Let us take the following numbers to be the zeros of the function we require:

$$z_t = e^{i\alpha_s}(1 - 2^{-k-s}), \ k \geqslant 1, \ s \geqslant 0.$$

Here, t is the serial number of a pair of numbers (k, s) ordered in some particular manner and

$$\alpha_s = \begin{cases} a_{\frac{s}{2}} & \text{for an even } s, \\ b_{\left[\frac{s}{2}\right]} & \text{for an odd } s. \end{cases}$$

It is easy to check that

$$\sum_{(t)} (1 - |z_t|) = 2 < \infty.$$

The required Blaschke function can now be written as an infinite product.

$$b(z) = \prod_t \frac{\bar{z}_t}{|z_t|} \cdot \frac{z_t - z}{1 - \bar{z}_t z}.$$

The infinite product converges absolutely and uniformly in any domain where $\inf_t |1 - \bar{z}_t z| = d > 0$, provided that several of the first factors are thrown away.

Let us obtain the estimate of the derivatives of b(z) that will be required below. First of all, let us note that for $|z| < 1$, $z \neq z_t$, we have

$$|[\ln b(z)]^{(r)}| = (r-1)! \left| \sum_t (z - z_t)^{-r} - (z - \bar{z}_t^{-1})^{-r} \right| \leqslant r! \sum_t |z - z_t|^{-r} |z - \bar{z}_t^{-1}|^{-r} \left| \int_{z_t}^{\bar{z}_t^{-1}} (z - \zeta)^{r-1} d\zeta \right| \leqslant$$

$$\leqslant \frac{r! \, 3^{r-1} \sum_t |z_t - \bar{z}_t^{-1}|}{\inf_t |z - z_t|^r \inf_t |z - \bar{z}_t^{-1}|^r} \leqslant \frac{r! \, 3^{r-1} \cdot 8}{\inf_t |z - z_t|^r \inf_t |z - \bar{z}_t^{-1}|^r}. \quad (2.19)$$

It is obvious that the following inequality holds:

$$b_{(z)}^{(r)} = \left[e^{\ln b(z)}\right]^{(r)} = b(z) \sum_{\sum n_i = r, \, n_i \geqslant 1} A_{\{n_i\}}^{(r)} \prod_i [\ln b(z)]^{(n_i)},$$

where the $A_{\{n_i\}}^{(r)}$ are integer-valued coefficients whose explicit form will not be given here.

Taking (2.19) into account, we find that for $|z| < 1$, $r \geq 1$, we have

$$|b^{(r)}(z)| \leqslant \frac{C(r)}{\inf_t |z - z_t|^r \inf_t |z - \bar{z}_t^{-1}|^r}. \quad (2.20)$$

Since the Blaschke function is bounded within the unit circle, we also have the usual estimate

$$|b^{(r)}(z)| \leqslant \frac{r! C_0}{\inf_{|\zeta|=1} |z - \zeta|^{r+1}}, \quad r \geqslant 1. \quad (2.21)$$

Combining (2.20) and (2.21), we obtain the bound on the derivatives of the Blaschke function. For this purpose, we will subdivide the unit circle into two sets X and Y. The set X will contain those points whose distance from the unit circle is less than the set of points z lying on the rays arg z = a_ν, arg z = b_ν. All other points on the unit circle will be put into set Y. The following inequalities can be easily verified:

$$\operatorname{dist}(z, \mathcal{E}) \leqslant \inf_{t,\tau}\{|z - z_t|, |z - \bar{z}_\tau^{-1}|\}, \ |z| \leqslant 1,$$
$$\frac{1}{3} \leqslant \frac{\operatorname{dist}(z, \mathcal{E})}{\operatorname{dist}(z, E)} \leqslant 1, \ x \in X,$$
$$1 \leqslant \frac{\operatorname{dist}(z, E)}{1 - |z|} \leqslant 3, \ x \in Y.$$

In view of what has been said above, the bounds (2.20) and (2.21) on X and Y can be replaced by

$$|b_{(z)}^{(r)}| \leqslant C_1(r)\{\operatorname{dist}(z, E)\}^{-2r}, \ r \geqslant 1, \ z \in X, \tag{2.22}$$

$$|b_{(z)}^{(r)}| \leqslant C_2(r)\{\operatorname{dist}(z, E)\}^{-r-1}, \ r \geqslant 1, \ z \in Y. \tag{2.23}$$

Combining (2.22) and (2.23), we directly obtain the required estimate for the derivatives of b(z), namely,

$$|b^{(r)}(z)| \leqslant D(r)\{\operatorname{dist}(z, E)\}^{-2r}, \ |z| < 1, \ r \geqslant 1. \tag{2.24}$$

In what follows, a significant part in what follows will be played by the function

$$F(z) = \frac{b(z)}{b(1)} \exp\{2^{\gamma+1}\pi^{-\gamma} - f(z)\}.$$

F(z) is regular in the unit circle and can be differentiated any number of times right up to the boundary: F(z) tends to zero together with all of its derivatives when z approaches the point z_0 belonging to **E** along any path lying within the unit circle. The proof of these assertions follows directly from (2.9), (2.18), and (2.24) if we make use of the fact that for any r > 0, we have

$$\rho^{-r} \exp\left[-\rho^{-\frac{\gamma}{3}}\right] \to 0 \quad \text{as} \quad \rho \to 0, \ \rho > 0.$$

Moreover, the equation F(z) = 0 has an infinite number of zeros within the unit circle accumulating to **E**. Let us also note that F(z) can be analytically continued to the exterior of the unit circle through the arcs L_ν.

With the help of formulas (2.7) and (2.8), we can easily establish that the following conditions are satisfied:

$$\sup_{0<\rho<1} |F(\rho e^{i\vartheta})| < \exp[h(0) - h(\vartheta)] < 1 \quad \text{for} \quad \vartheta \neq 0,$$
$$\lim_{\rho\to 1} F(\rho) = 1. \tag{2.25}$$

We will now require the following simple assertion:

LEMMA 1. *Let g(z) be a function regular in the circle |z| < 1, bounded right up to the contour, and continuable across an arc of the unit circle containing the point z = 1. If the conditions*

$$|g(z)| < 1 \quad \text{for} \quad |z| < 1,$$
$$\lim_{\rho\to 1} g(\rho) = 1$$

are satisfied, then we have

$$g'(1) \neq 0.$$

PROOF. Let us show that the assumption that we have $g'(1) = 0$ leads to a contradiction.

We will assume at first that $g''(1) = \rho e^{i\varphi} \neq 0$, where $|\varphi| \leq \pi$. Then, in view of the regularity of $g(z)$ for $|z-1| \ll 1$, we can write

$$g(z) = 1 + \frac{1}{2!}\rho |z-1|^2 e^{i\{\varphi + 2\arg(z-1)\}} + O(|z-1|^3).$$

It is obvious that we can take two positive numbers ε and α to be so small that the point

$$z_0 = 1 + \varepsilon \exp i\left\{\pi - \frac{\varphi}{2} + \alpha \operatorname{sgn} \varphi\right\}$$

belongs to the circle $|z| < 1$. Calculating the value of $g(z)$ at this point, we find that for small ε and α we have

$$|g(z_0)| \geqslant 1 + \frac{\varepsilon^2}{2}\rho \cos\alpha - \frac{\varepsilon^2}{2}\rho |\sin\alpha| + O(\varepsilon^3) > 1,$$

which is in contradiction to the condition of the lemma.

Let us now take $g'(1) = g''(1) = \ldots = g^{r-1}(1) = 0$ and $g^{(r)}(1) = \rho e^{i\varphi} \neq 0$, where $|\varphi| \leq \pi$, $r \geq 3$. For any integral $r \geq 3$ we can find at least one complex number α_r that is the r-th root of unity and that satisfies the condition

$$\frac{\pi}{2} < -\frac{\varphi}{r} + \arg \alpha_r < \frac{3\pi}{2}.$$

Then, the point

$$z_0 = 1 + \varepsilon \alpha_r e^{-\frac{i\varphi}{r}}$$

for small ε will be inside the circle $|z| < 1$; calculating the value of $g(z)$ at this point, we obtain

$$|g(z_0)| = \left|1 + \frac{1}{r!}\varepsilon^r + O(\varepsilon^{r+1})\right| > 1,$$

which contradicts the condition of the lemma. The lemma has thus been proved.

It follows directly from Lemma 1 that $F'(1) \neq 0$ from which we see that $F(z)$ is a one-sheeted function in a small neighborhood of the point $z = 1$.

We will need to make a more detailed study of the set of values of $F(z)$. We will base our investigation on the following lemma.

LEMMA 2. Let $g(z)$ be a function that is regular in the circle $|z| < 1$, bounded right up to the contour, analytically continuable across an arc of the unit circle $z = e^{i\vartheta}$, where $a_0 < \vartheta < b_0$, $(a_0 \cdot b_0 < 0)$, and let the following conditions be satisfied:

$$\begin{aligned} &\frac{d}{d\vartheta}|g(e^{i\vartheta})| > 0 \quad \text{when} \quad \vartheta \in (a_0, 0), \\ &\frac{d}{d\vartheta}|g(e^{i\vartheta})| < 0 \quad \text{when} \quad \vartheta \in (0, b_0). \end{aligned} \tag{2.26}$$

Moreover, let us take

$$\sup_{0\leqslant\rho<1} |g(\rho e^{i\vartheta})| < 1, \ \vartheta \neq 0,$$
$$\lim_{\rho\to 1} g(\rho) = 1.$$

Then, (1) the boundary of the value set $R(g)$ of $g(z)\ |z| < 1$, contains an analytic arc Λ which is the image of some arc $L(a_0, b_0)$, $a_0 b_0 < 0$, of the unit circle, namely,

$$g(z) = \zeta \in \Lambda \text{ for } z \subset L;$$

and (2) the set $R(g)$ is contained in a closed domain bounded by an analytic arc Λ and an arc of the circle $|\xi| = 1 - \varepsilon$, where $\varepsilon > 0$.

PROOF. In view of Lemma 1, we have $g'(1) \neq 0$ and, consequently, $g(z)$ is single-sheeted in a domain ω_δ which is the intersection of the δ-neighborhood of the point $z = 1$, $\delta \ll 1$, and the unit circle $|z| < 1$. Let us draw the level line on which $|g(z)| = 1 - \varepsilon$. For all sufficiently small ε this line is a simple analytic arc situated completely in ω_δ. Indeed, there would otherwise be a point $z_1 \neq 1$, $|z_1| < 1$ such that $|g(z_1)| = 1$ and this contradicts the conditions of the lemma. Also, it is not difficult to see that the level line $|g(z)| = 1 - \varepsilon$ does not intersect itself and intersects the arc (a_0, b_0) at only two points $\alpha \in (a_0, 0)$ and $\beta \in (0, b_0)$. The first assertion is a consequence of the fact that $\ln|g(z)|$ is harmonic in the neighborhood of the point $z = 1$, the second can be directly derived from (2.26). Let us denote the arc (α, β) of the unit circle by L. The domain ω bounded by the level line $|g(z)| = 1 - \varepsilon$ and the arc L satisfies the following requirements as can be easily established:

1. For each $z_0 \in \overline{\omega}$, the equation $g(z) = g(z_0)$ has a single root $z = z_0$ in the circle $|z| \leq 1$.

2. If $z \in \overline{\omega}$, $z_1 \overline{\in} \overline{\omega}$, $|z_1| \leq 1$, then the inequality $|g(z)| \geq 1 - \varepsilon > |g(z_1)|$ is satisfied. It follows directly from this that the value set $R(g)$ of $g(z)$, $|z| \leq 1$ lies inside the domain bounded by an arc of the circle $|\xi| = 1 - \varepsilon$ and an arc Λ which is the image of L. In view of the analyticity of $g(z)$ on L, the arc Λ is also an analytic arc. The lemma has been proved.

REMARK. It is not difficult to construct a domain $\Omega_0 \supset \Omega$ with an infinitely differentiable boundary containing a part Λ_0 of arc Λ. Then, we can assert that $g(z)$ maps the unit circle into the interior of the domain Ω_0, an arc $N_0 \subset L$ of the unit circle being mapped onto the analytic arc Λ_0 which is part of the boundary of Ω_0.

It is easy to see that all the conditions of Lemma 2 are satisfied for the function $F(z)$ constructed above if we take $a_0 = -\pi/2$, $b_0 = \pi/2$. Consequently, $F(z)$ maps the unit circle into the interior of a domain Ω_0 with an infinitely differentiable boundary which contains an analytic arc Λ_0 that is the F image of an arc $N_0(\alpha_0, \beta_0)$ of the unit circle $z = e^{i\vartheta}$, where $-\pi/2 < \alpha_0 < \vartheta < \beta_0 < \pi/2$, $\alpha_0 \cdot \beta_0 < 0$.

Let us introduce the function $\varphi(\xi)$ which represents the conformal mapping of Ω_0 onto the unit circle in such a manner that the arc Λ_0 becomes the arc L_0 of the unit circle. In view of well-known theorems (for example, see [8]), $\varphi(\xi)$ is an infinitely differentiable function in $\overline{\Omega}_0$. Consequently, the function

$$\varphi[F(z)] \equiv \Phi(z)$$

is analytic in the unit circle and is continuous right up to the boundary together with all of its derivatives. Moreover, we have $|\Phi(z)| \leq 1$ and the point $\Phi(e^{i\vartheta})$ lies on the arc L_0 of the unit circle when $\vartheta \in \alpha_0, \beta_0$. It should also be noted that the equation $\Phi(z) = \varphi(0)$ has an infinite number of roots in the unit circle and these roots accumulate to the set **E** situated on the unit circle.

Let us use $\Phi(z)$ to construct the function $m(k)$ which satisfies the conditions of the Principal Lemma. We carry out several conformal transformations for this purpose.

Let us subdivide the arc N_0 of the unit circle into three parts, i.e., $N_0 = N_0' + N_0'' + N_0'''$. The images of N_0', N_0'', and N_0''' in the value set of $\Phi(z)$ are the arcs L_0', L_0'', and L_0''', where $L = L_0' + L_0'' + L_0'''$. Let $\tilde{\psi}(z)$ denote the function that maps the unit circle $|z| < 1$ onto the semicircle $\operatorname{Im} z > 0$, $|z| < 1$ in such a manner that the arc N_0'' becomes the diameter $(-1, 1)$ and the arcs N_0', N_0'' become the contiguous arcs s', s'' of the unit circle. Next, let $\psi(z)$ denote the function that maps the unit circle onto the semicircle $\operatorname{Im} z > 0$, $|z| < 1$ in such a manner that the arc L_0'' becomes the diameter $(-1, 1)$ and L_0', L_0''' become the contiguous arcs S', S'' of the unit circle. $\tilde{\psi}(z)$ and $\psi(z)$ can be easily expressed in terms of elementary functions. Using $\tilde{\psi}_1(z)$ to denote the function that is the inverse of $\tilde{\psi}(z)$, we can form the composition

$$\Psi(z) = \psi\{\Phi[\tilde{\psi}_1(z)]\}.$$

It is not difficult to see that this function is defined in the semicircle $|z| \leq 1$, $\operatorname{Im} z \geq 0$, and maps it into itself, the diameter $(-1, 1)$ being mapped into the diameter $(-1, 1)$, while the contiguous arcs s', s'' of the unit circle are mapped into the arcs S', S''. Making use of the principle of symmetry, we continue $\psi(z)$ into the lower semicircle. The function obtained as the result of the continuation will be denoted by $\chi(z)$. Obviously, it satisfies the conditions

$$|\chi(z)| < 1 \quad \text{for} \quad |z| < 1,$$
$$\operatorname{Im} \chi(z) \operatorname{Im} z > 0.$$

It should be noted that $\chi(z)$ maps the arcs $\bar{s}'$, $\bar{s}''$ of the lower semicircle that are symmetrical to the arcs s', s'' into the arcs $\tilde{S}'$, $\tilde{S}''$ of the lower semicircle that are symmetrical to S', S''.

Now, once again using the symmetry principle, we continue $\chi(z)$ across the arc $s' + \bar{s}'$ and $s'' + \bar{s}''$ into the exterior of the unit circle. The function obtained in this way will be denoted by $G(z)$. It is regular inside the unit circle and in the vicinity of the points $z = \pm 1$. It follows from the properties of $\Psi(z)$ that $G(z)$ is infinitely differentiable right up to the unit circle and that

$$|G(z)| < 1 \quad \text{for} \quad |z| < 1.$$

On the basis of Lemma 1 we can assert that $G'(1) \neq 0$ and $G'(-1) \neq 0$. With the help of the boundary-correspondence principle, this directly yields

$$G'(1) > 0, \quad G'(-1) > 0.$$

In view of (2.26), the function $G(z)$ satisfies the inequality

$$\operatorname{Im} G(z) \operatorname{Im} z > 0 \quad \text{for} \quad |z| < 1. \tag{2.27}$$

It is also not difficult to see that the equations

$$G(z) = \psi[\varphi(0)],$$
$$G(z) = \overline{\psi[\varphi(0)]}$$

have an infinite number of roots in the upper and lower semicircles, respectively. The roots of these equations accumulate to points which form the sets

$$\Delta_E \left\{z : z = \widetilde{\psi}(e^{i\theta}),\quad e^{i\theta} \in E\right\}$$
$$\Delta'_E \left\{z : z = \overline{\widetilde{\psi}(e^{i\theta})},\quad e^{i\theta} \in E\right\},$$

respectively. The sets Δ_E and Δ'_E are obviously similar to the set **E** in the sense of the definition given at the beginning of this section.

Let us now consider the function

$$m(k) = [2G'(1)]^{-1}\left\{G\left(\frac{k-i}{k+i}\right) - 1\right\} \tag{2.28}$$

and let us show that it satisfies all of the conditions of the Principal Lemma.

(1) The function m(k) is regular in the half-plane Im $k > 0$ and is continuous together with all of its derivatives right up to the real axis.

(2) $m(0) = -[G'(1)]^{-1} \neq 0$, $m'(0) = -[G'(1)]^{-1}G'(-1) \neq 0$.

(3) m(k) is regular near the points $k = 0$, $k = \infty$, which follows from the regularity of G(z) at the points $z = \pm 1$. Moreover, we have $km(k) = k[2G'(1)]^{-1}\left\{G\left(\frac{k-i}{k+i}\right) - 1\right\} \to -i$ as $k \to \infty$.

(4) m(k) assumes real values on the positive imaginary axis. Indeed, assuming that $k = i\varkappa$, $\varkappa > 0$, we have

$$\operatorname{Im} m(i\varkappa) = [2G'(1)]^{-1} \quad \operatorname{Im} G\left(\frac{\varkappa - 1}{\varkappa + 1}\right) = 0$$

since $-1 < (\varkappa - 1)(\varkappa + 1)^{-1} < 1$ for $\varkappa > 0$. Let us verify that Im $(-k) = -\operatorname{Im} m(k) > 0$ for $k > 0$ with the help of the symmetry principle

$$\operatorname{Im} m(-k) = [2G'(1)]^{-1} \operatorname{Im} G\left(\frac{-k-i}{-k+i}\right) = [2G'(1)]^{-1} \operatorname{Im} G\left(\frac{k-i}{k+i}\right) =$$
$$= -[2G'(1)]^{-1} \operatorname{Im} G\left(\frac{k-i}{k+i}\right) = -\operatorname{Im} m(k).$$

That Im $m(-k)$ is positive for $k > 0$ follows from (2.27).

Thus, all the conditions of the Principal Lemma are satisfied.

Consequently, there exists an infinitely differentiable real function q(x) with $q(x) \in S_\infty$ and a real number a_0 such that*

$$-[m(\sqrt{\lambda})]^{-1} + a_0$$

is a Weyl function of the differential operator generated in $L_2(0, \infty)$ by the differential expression

$$ly = -y'' + q(x)y \tag{2.29}$$

and the boundary condition $y(0) = 0$. Introducing the abbreviations

$$h_1 = 2G'(1)\{\psi[\varphi(0)] - 1\}^{-1} - a_0,$$
$$h_2 = \overline{h_1}$$

and making use of formula (1.5), we conclude that the operators l_{h_1}, $l_{h_2} = l^*_{h_1}$ generated in L_2 $(0, \infty)$ by the differential expression (2.29) and the boundary condition

* As usual, the appropriate branch of the function $\sqrt{\lambda}$ is selected by means of the condition Im $\sqrt{\lambda} > 0$.

$$y'(0) - h_1 y(0) = 0 \quad \text{or} \quad y'(0) - h_2 y(0) = 0$$

have an infinite number of eigenvalues.

The set of the eigenvalues of the operator l_{h_1} (l_{h_2}) is infinite and is situated in the upper (lower) half-plane. The points to which the eigenvalues accumulate fill the set $\mathbf{E}_3$ lying on the positive real semi-axis $\lambda > 0$. The set $\mathbf{E}_3$ is obviously similar to $\mathbf{E}$ and coincides with the set of the singular points of infinite multiplicity of the operator l_h for $h = h_1, h_2$. This completes proof of the theorem.

REMARK 1. For all other values of h, the operator l_h has no singular points of infinite multiplicity and, consequently, the number of eigenvalues of l_h is finite.

REMARK 2. In constructing the function m(k), we have used the Blaschke function all of whose roots are simple. It is not difficult to see that because of this all roots of the equation

$$[m(\sqrt{\lambda})]^{-1} - a_0 = h_1$$

are also simple and the eigenvalues of the operator l are of unit rank.

If we replace the function b(z) by

$$\tilde{b}(z) = b(z)(z - z_0)^m (z - \bar{z}_0^{-1})^{-m},$$

then one of the eigenvalues of l_{h_1} will be of rank m.

§ 3. The Operator with Potential $q(x) \in S_n$, $n < \infty$

It has been shown in [1, 2] that the accumulation-point set of the eigenvalues of an operator l_h with a complex potential $q(x) \in S_n$, $n \leq 1$ is of measure zero and satisfies the condition

$$\Sigma |l_\nu| |\ln |l_\nu|| < \infty. \tag{3.30}$$

Here, $|l_\nu|$ is the length of the interval l_ν of contiguity with the set of eigenvalue-accumulation points and the summation extends over all bounded contiguity intervals.

In the present section we will show that the condition (3.30) is not an overestimate. More accurately, we will prove the following assertion.

THEOREM II. *Let* $\mathbf{E}$ *be an arbitrary bounded closed point set of the real axis which is of zero measure and which satisfies condition (3.30) and let n be an arbitrary integer,* $n \geq 2$.

Then, there exists a differential operator of the form of l *with an infinitely differentiable real potential* $q(x) \in S_n$ *such that for some boundary condition* $y'(0) - hy(0) = 0$, $\operatorname{Im} h \neq 0$, *the set of its eigenvalue-accumulation points is similar* to* $\mathbf{E}$.

PROOF. The proof of Theorem II consists in the construction of a function $m_1(k)$ which is analytic in the half-plane Im k > 0 and which satisfies the conditions of the Principal Lemma for the given value of n. In addition, we will require that the equation

*Here and in the following, similarity between two sets is to be understood in the sense of the definition given in the preceding section.

$$m_1(k) = a$$

for some value of a has an infinite number of roots in the upper half-plane and that they accumulate to a point set Δ_1 on the real half-axis $k > 0$. We will carry out this construction in such a manner that this point set will be similar to the set* E.

The proof of Theorem II differs from that of Theorem I only in the use of a different auxiliary function $F_1(z)$. The second part of the proof [the construction of m(k) from the function F(z) with known properties] is repeated almost completely, the only difference being that the analogs of $\Phi(z)$ and $G(z)$ are regular in the circle $|z| < 1$, but only have n derivatives that are continuous right up to the boundary of the circle.

Let us now construct the auxiliary function $F_1(z)$.

It is obvious that we can assume that the set **E** is situated on the segment $-\pi/2 \le \vartheta \le 3\pi/2$ and that the points $\pi/2$, $3\pi/2$ belong to **E**. Let a_ν and b_ν denote the end points of bounded intervals of contiguity with the set **E**. Having set $a_0 = -\pi/2$, $b_0 = \pi/2$, let us map **E** onto the unit circle and let us introduce the function†

$$h(\vartheta) = \begin{cases} \ln \dfrac{\pi}{\vartheta - a_\nu} + \ln \dfrac{\pi}{b_\nu - \vartheta} \equiv h_\nu(\vartheta), & \vartheta \in (a_\nu, b_\nu), \ \nu \geqslant 0 \\ +\infty \text{ for all other values of } \vartheta \in \left(-\frac{\pi}{2}, \frac{3\pi}{2}\right]. \end{cases} \tag{3.31}$$

The function $h(\vartheta)$ is analytic on the intervals (a_ν, b_ν), $\nu \ge 0$, and is summable over the unit circle as follows from the equality

$$\int_{-\frac{\pi}{2}}^{\frac{3\pi}{2}} |h(\vartheta)|\, d\vartheta = \sum_{\nu=0}^{\infty} \int_{a_\nu}^{b_\nu} h_\nu(\vartheta)\, d\vartheta = \sum_{\nu=0}^{\infty} 2(b_\nu - a_\nu) \ln \frac{\pi}{b_\nu - a_\nu} + 4\pi. \tag{3.32}$$

The Poisson integral

$$u(r, \varphi) = \frac{1}{2\pi} \int_{-\frac{\pi}{2}}^{\frac{3\pi}{2}} \frac{(1 - r^2)\, h(\vartheta)}{1 + r^2 - 2r\cos(\varphi - \vartheta)}\, d\vartheta, \quad r < 1, \quad -\frac{\pi}{2} < \varphi \leqslant \frac{3\pi}{2},$$

yields a function that is harmonic in the unit circle. From the maximum principle it follows that the following inequality holds at all internal points of the unit circle:

$$u(r, \varphi) > \inf h(\vartheta) = h(0) = 2\ln 2. \tag{3.33}$$

---

* The properties of $m_1(k)$ show that the equations $m_1(k) = a$ and $m_1(k) = \bar{a}$ have the same number of roots. If the roots of the first equation accumulate to the set Δ_1 situated on the half-axis $k > 0$, then the roots of the second equation accumulate to the set Δ_1' situated on the half-axis $k < 0$ symmetrically to Δ_1.

† An analogous function has been used by L. Carleson in the proof of one uniqueness theorem for functions that are regular in the unit circle (see [9]).

Repeating the arguments given at the beginning of the last section, we can easily find, on the basis of the obvious bound,

$$h(\vartheta) > \ln \frac{\pi}{\inf\limits_{\vartheta_0 \in E} |\vartheta - \vartheta_0|}, \tag{3.34}$$

that the lower bound to the function $u(r, \varphi)$ in the unit circle

$$u(r, \varphi) > K_0 \ln \frac{1}{\operatorname{dist}[(r, \varphi), E]}.$$

Here, as is usual, dist $[(r, \varphi), \mathbf{E}]$ denotes the distance from the point with coordinates (r, φ) to the set $\mathbf{E}$; K_0 is a positive constant whose exact value does not interest us.

Let us now introduce a function that is analytic in the unit circle and whose real part coincides with $u(r, \varphi)$, namely,

$$f(z) = \frac{1}{2\pi} \int_{-\frac{\pi}{2}}^{\frac{3\pi}{2}} \frac{e^{i\vartheta} + z}{e^{i\vartheta} - z} h(\vartheta)\, d\vartheta, \quad |z| < 1. \tag{3.35}$$

Let us find the estimates of the derivatives of $f(z)$. For this purpose, making use of definition (3.31) of $h(\vartheta)$, we will rewrite (3.35) as

$$f(z) = \sum_{\nu=0}^{\infty} f_\nu(z), \tag{3.36}$$

where

$$f_\nu(z) = \frac{1}{2\pi i} \int_{L_\nu} \frac{x + z}{x - z} h_\nu\left(\frac{1}{i} \ln x\right) \frac{dx}{x}. \tag{3.37}$$

Here, L_ν is an arc of the unit circle $x = e^{i\varphi}$, $\vartheta \in (a_\nu, b_\nu)$. The appropriate branch of ln x should be selected on the basis of the condition $\ln e^{ia_0} = ia_0$. In view of (3.32), the series (3.36) converges absolutely and uniformly in any circle $|z| \le 1 - \delta$, $\delta > 0$. It should be noted that the function

$$\tilde{h}_\nu(x) = x^{-1} h_\nu\left(\frac{1}{i} \ln x\right)$$

becomes singly-valued and regular in domain $|x| > 1$, $a_\nu < \arg x < b_\nu$ after we choose the branch of h_ν from the condition

$$\tilde{h}_\nu\left(e^{i\frac{b_\nu - a_\nu}{2}}\right) = 2 \ln \frac{2\pi}{b_\nu - a_\nu} e^{i\frac{b_\nu + a_\nu}{2}}$$

and after we introduce the cuts along $\arg x = a_\nu$, $\arg x = b_\nu$. We will show that $\tilde{h}_\nu(x)$ is summable over the segments Γ_ν^1 and Γ_ν^3 of the rays $\arg x = a_\nu$, $\arg x = b_\nu$, $1 < |x| < \exp(b_\nu - a_\nu)$, and over the arc Γ_ν^2 of the circle $|x| = \exp(b_\nu - a_\nu)$, $a_\nu < \arg x < b_\nu$. Let us evaluate the integral over Γ_ν^1

$$\int_{\Gamma_\nu^1} |\tilde{h}_\nu(x)|\, d|x| = \int_1^{\exp(b_\nu - a_\nu)} \frac{1}{|x|} \left| \ln \frac{\pi}{\frac{1}{i} \ln|x|} + \ln \frac{\pi}{\frac{1}{i} \ln|x| + b_\nu - a_\nu} \right| d|x| \leqslant$$

$$\leqslant 2(b_\nu - a_\nu)|\ln(b_\nu - a_\nu)| + (b_\nu - a_\nu)(1 + \ln \pi + \pi). \tag{3.38}$$

This is also the bound on the integral

$$\int_{\Gamma_\nu^3} |\tilde{h}_\nu(x)|\, d|x|.$$

Further, we have

$$\int_{\Gamma_\nu^2} |\tilde{h}_\nu(x)|\,|x|\, d\arg x \leqslant 2\int_{a_\nu}^{b_\nu} \left|\ln \frac{\pi}{\frac{1}{i}(b_\nu - a_\nu) + t - a_\nu}\right| dt \leqslant$$

$$\leqslant 2 \int_0^{b_\nu - a_\nu} \ln\frac{\pi}{t}\, dt + \pi(b_\nu - a_\nu) = (b_\nu - a_\nu)\,[2|\ln(b_\nu - a_\nu)| + 1 + \pi + \ln\pi]. \tag{3.39}$$

Integral (3.37) may be replaced by an integral over the contour $\Gamma_\nu = \Gamma_\nu^1 + \Gamma_\nu^2 + \Gamma_\nu^3$,

$$f_\nu(z) = \frac{1}{2\pi i}\int_{\Gamma_\nu} \frac{x+z}{x-z}\tilde{h}_\nu(x)\,dx, \quad |z| < 1,$$

where we have

$$\int_{\Gamma_\nu} |h_\nu(x)|\, d\Gamma_\nu \leqslant 3(b_\nu - a_\nu)\,[2|\ln(b_\nu - a_\nu)| + 1 + \pi + \ln\pi].$$

This immediately yields

$$\begin{aligned} |f_\nu(z)| &\leqslant \frac{3}{2\pi}\left[\frac{2(1+e^\pi)}{\operatorname{dist}(z, \Gamma_\nu)} + 1\right](b_\nu - a_\nu)\,[2|\ln(b_\nu - a_\nu)| + 1 + \pi + \ln\pi], \\ |f_\nu^{(r)}(z)| &< \frac{3(1+e^\pi)\, r!}{\{\operatorname{dist}(z, \Gamma_\nu)\}^{r+1}}(b_\nu - a_\nu)\,[2|\ln(b_\nu - a_\nu)| + 1 + \pi + \ln\pi], \\ & \qquad r \geqslant 1. \end{aligned} \tag{3.40}$$

Noting that

$$\inf_\nu\, [\operatorname{dist}(z, \Gamma_\nu)] = \operatorname{dist}(z, E), \quad |z| \leqslant 1, \tag{3.41}$$

and summing (3.40) over ν, we find that in view of (3.36) we have

$$|f^{(r)}(z)| < \frac{r!}{\{\operatorname{dist}(z, E)\}^{r+1}}\, 6\left[\sum_\nu (b_\nu - a_\nu)|\ln(b_\nu - a_\nu)| + \pi(1 + \pi + \ln\pi)\right].$$

In the same way as in the preceding section we can construct the Blaschke function $b_1(z)$ whose roots z_t accumulate to E. The derivatives of $b_1(z)$ have the bounds

$$|b_1^{(r)}(z)| < K_r\{\operatorname{dist}(z, E)\}^{-2r}, \quad |z| < 1, \quad r \geqslant 1. \tag{3.42}$$

We now introduce the auxiliary function

$$F_1(z) = [b_1(1)]^{-1}\, b_1(z) \exp N\{2\ln 2 - f(z)\},$$

where $N = 2K_0^{-1}(n+1)$ and K_0 is the same constant as in formula (3.34).

$F_1(z)$ is regular in the unit circle, analytically continuable across arcs L_ν into the exterior of the circle, has roots which accumulate to E at the points z_t, and satisfies the conditions

$$\sup_{0\leq\rho<1} |F_1(\rho e^{i\vartheta})| < 1 \quad \text{for} \quad \vartheta \neq 0,$$
$$\lim_{\rho\to 1} F_1(\rho) = 1.$$

Using (3.34), (3.41), (3.42), we can easily obtain the bound on the derivatives of $F_1(z)$ in the circle $|z| < 1$

$$|F_1^{(r)}(z)| \leqslant C(r)\{\text{dist}(z, E)\}^{-2r} \exp\{2(n+1)\ln \text{dist}(z, E)\} = C(r)\{\text{dist}(z, E)\}^{2(n-r+1)}, \quad r=0, 1, 2, \ldots \tag{3.43}$$

Since $F_1(z)$ is regular within the unit circle and is analytically continuable across arcs L_ν, $\nu \geq 0$, we conclude that $F_1(z)$ is continuous right up to the circle together with all of its derivatives up to and including the n-th derivative and

$$|F_1^{(r)}(z)| \to 0 \quad \text{as} \quad z \to e^{i\vartheta_0}, \ \vartheta_0 \in E, \ r=0, 1, \ldots n.$$

This concludes the construction of the auxiliary function $F_1(z)$. As was noted earlier, the remaining part of the proof is a repetition of the corresponding part of the proof of Theorem I, the only difference being that we are now not dealing with an infinitely differentiable function, but with a function which is continuous in the unit circle together with its first n derivatives. In view of this, we will not give this part of the proof here.

The next theorem shows that the set of eigenvalue-accumulation points may have a positive measure if the potential q(x) of operator l_h does not decrease at infinity.

THEOREM III. There exists a differential operator of the form of l_h with an infinitely differentiable real potential such that with a complex boundary condition

$$y'(0) - hy(0) = 0, \quad \text{Im}\, h \neq 0,$$

the set of eigenvalue-accumulation points fills a segment of the positive half-axis $\lambda > 0$.

PROOF. The proof of this theorem, in the same way as the proofs of Theorems I and II, consists in the construction of the Weyl function of the required operator.

Let us first of all construct an auxiliary function $F_2(z)$.

Let us consider the Blaschke function with roots at the points

$$z_{mn} = (1 - 2^{-m-n}) \exp i\left\{\frac{\pi}{2} + \frac{m}{n}\pi\right\},$$

where m, n is a pair of natural numbers, m < n; the function is

$$b_2(z) = \prod_{m<n} \frac{z_{mn} - z}{1 - \bar{z}_{mn} z} \frac{\bar{z}_{mn}}{|z_{mn}|}. \tag{3.44}$$

The infinite product (3.44) defines a function which is regular in the unit circle, and which is analytically continuable across an arc of the unit circle $z = e^{i\vartheta}$, $-\pi/2 < \vartheta < \pi/2$. The zeros of $b_2(z)$ accumulate to each point of the semicircle $z = e^{i\vartheta}$, $\pi/2 \leq \vartheta \leq 3\pi/2$.

Let us now consider the auxiliary function

$$F_2(z) = [b_2(1)]^{-1} b_2(z) \exp\{2\ln 2 - f(z)\},$$

where $f(z)$ is the function defined by formula (3.35).

The function $F_2(z)$ is regular in the circle $|z| < 1$ and is analytically continuable across the arc of $z = e^{i\vartheta}$, $-\pi/2 < \vartheta < \pi/2$. Further, it is not difficult to establish that the following conditions hold:

$$\sup_{0<\rho<1} |F_2(\rho e^{i\vartheta})| < 1, \quad \vartheta \neq 0,$$
$$\lim_{\rho \to 1} F_2(\rho) = 1.$$

We can construct $m_2(k)$ from the auxiliary function $F_2(z)$ in the same way as we have constructed $m(k)$ from $F(z)$ in the preceding section. It is not difficult to see that $m_2(k)$ constructed in this manner is regular in the half-plane Im $k > 0$, real on the positive imaginary half-axis, has a negative imaginary part in the first quadrant of the k plane and a positive imaginary part in the second quadrant, and satisfies the conditions 1 and 2 of the Principal Lemma. Moreover, for some $\varkappa$ with Im $\varkappa < 0$, the equation

$$m_2(k) = \varkappa \tag{3.45}$$

has an infinite number of roots in the first quadrant accumulating to a segment of the real axis.

It should be noted that by contrast with the preceding theorems, the function $m_2(k)$ is no longer continuous right up to the real axis.

Let us consider the nondecreasing function*

$$\rho_\infty(\lambda) = \begin{cases} \lim\limits_{v \to +0} \int\limits_0^\lambda \operatorname{Im} m_2^{-1}(\sqrt{u+iv})\, du, & \lambda > 0, \\ 0, & \lambda \leqslant 0. \end{cases} \tag{3.46}$$

From the properties of $m_2(k)$ listed above, we find in the same manner as in the proof of the Principal Lemma that there exists a unique differential operator of the form of l_∞ with an infinitely differentiable potential $q(x)$ and that $\rho_\infty(\lambda)$ is the spectral function of this operator [with the boundary condition $y(0) = 0$].

It should be noted that the absence of smoothness in the function $\rho_\infty(\lambda)$ means that we cannot determine the behavior of $q(x)$ at infinity. We can only assert that since the operator constructed is semi-bounded, we have the case of a limit point (see [4]).

It can be seen from (3.46) that the Weyl function $m_\infty(\lambda)$ operator we have constructed differs from the function $-[m_2(\sqrt{\lambda})]^{-1}$ merely by a first-degree polynomial, i.e.,

$$m_\infty(\lambda) = -[m_2(\sqrt{\lambda})]^{-1} + b\lambda + a.$$

Here, we have $b \leq 0$ and a is a real number. On the basis of well-known formulas (see [4]), this yields the following expression for the spectral functions of the operator l_0 with potential $q(x)$ and the boundary condition $y'(0) = 0$:

$$\rho_0(\lambda) = \lim_{v \to +0} \int_0^\lambda \operatorname{Im}\{-[b(u+iv) + a - (m_2(\sqrt{u+iv}))^{-1}]^{-1}\}\, du.$$

Taking into account that $m_2(k)$ satisfies Condition 3 of the Principal Lemma and making use of the well-known asymptotic expression for the spectral function $\rho_0(\lambda)$ (see [10]) for $\lambda \to \infty$,

* As usual, the appropriate branch of the function $\sqrt{\lambda}$ is selected from the condition Im $\sqrt{\lambda} > 0$ for all nonpositive nonzero λ.

namely,

$$p_0(\lambda) = \sqrt{\lambda} + o(1),$$

we find that b = 0 and, consequently, we have

$$m_\infty(\lambda) = a - \left[m_2(\sqrt{\lambda})\right]^{-1}.$$

Taking into account what has been said above concerning the roots of Eq. (3.45) and making use of formula (1.5), we find that the differential operator in question has an infinite number of eigenvalues when the boundary conditions are

$$y'(0) - \left(\frac{1}{\varkappa} - a\right) y(0) = 0,$$

$$y'(0) - \left(\frac{1}{\bar{\varkappa}} - a\right) y(0) = 0.$$

The set of eigenvalue-accumulation points is obviously a segment and the theorem has been proved.

LITERATURE CITED

1. B. S. Pavlov, "The spectral theory of nonself-adjoint operators," Dokl. Akad. Nauk SSSR, Vol. 146, No. 6 (1962).
2. B. S. Pavlov, "The nonself-adjoint Schroedinger operator. II," in: Topics in Mathematical Physics, Vol. 2, Consultants Bureau, New York (1968).
3. H. Weyl, "Über gewöhnliche Differentialgleichungen mit Singularitäten und die zugehörigen Entwicklungen willkürlicher Functionen," Math. Ann., Vol. 68 (1910).
4. E. C. Titchmarsh, Eigenfunction Expansions Associated with Second-Order Differential Equations, Vol. I, Oxford University Press (1962).
5. V. A. Marchenko, "Expansions in eigenfunctions of nonself-adjoint second-order singular differential operators," Matem. Sborn., 52(94):2 (1960).
6. B. S. Pavlov, "The nonself-adjoint Schroedinger operator," in: Topics in Mathematical Physics, Vol. 1, Consultants Bureau, New York (1967).
7. I. I. Privalov, The Boundary Properties of Analytic Functions, GITTL (1950).
8. S. Warschawski, "On the differentiability at the boundary in conformal mapping," Proc. Am. Math. Soc., Vol. 12, No. 4 (1961).
9. L. Carleson, "Sets of uniqueness for functions regular in the unit circle," Acta Math., Vol. 87, No. 3-4 (1952).
10. I. M. Gel'fand and B. M. Levitan, "The construction of a differential equation from its spectral function," Izv. Akad. Nauk SSSR, Ser. Math., Vol. 15 (1951).

THE SINGULAR NUMBERS OF THE SUM OF COMPLETELY CONTINUOUS OPERATORS

S. Yu. Rotfel'd

The relation between the singular numbers (s-numbers) of a sum of completely continuous operators and the singular numbers of the individual terms has been studied in [1-4]. In particular, the results of [1] allow us to introduce a symmetric norm (see [5]) in some ideals of the ring **R** of all bounded linear operators acting in Hilbert space.

The present paper contains some new relations between the s-numbers of the sum of linear operators and these supplement the well-known inequalities of Fan Ky, Lidskii and Wielandt, and Amir-Moez. On the basis of these relations it is possible to introduce a unitary-invariant metric in nonnormable ideals of ring **R**. The results obtained here are then applied to the theory of Stieltjes double-integral operators.

The results presented below have been previously communicated without proof in [9].

1. Let **H** be a separable Hilbert space and let $\mathbf{S}_\infty$ be the set of all completely continuous operators acting in **H**. If we have $A \in \mathbf{S}_\infty$, then we use $\{s_k(A)\}_1^\infty$ to denote the sequence of the singular numbers of operator A (see [5]), i.e., the sequence of the eigenvalues $\{\lambda_k(|A|)\}_1^\infty$ of the operator $|A| = (A^*A)^{1/2}$ arranged in nonincreasing order with multiplicity taken into account. As is well known, any operator A belonging to $\mathbf{S}_\infty$ can be represented as

$$A = \sum_k s_k(A)(\cdot, \omega_k)\theta_k,$$

where $\{\omega_k\}$, $\{\theta_k\}$ are two orthonormal systems.

The principal result of this paper is the following theorem:

THEOREM 1. *Let $f(x)$, where $0 \le x < +\infty$ be a concave function such that $f(0) = f(+0) = 0$.*

If we have $A, B \in S_\infty$ and $C = A + B$, then for any natural number n the following inequality is satisfied:

$$\sum_{k=1}^{n} f(s_k(C)) \leqslant \sum_{k=1}^{n} f(s_k(A)) + \sum_{k=1}^{n} f(s_k(B)). \tag{1}$$

The proof of the theorem is given in Section 3.

2. Let us examine the finite-dimensional case first.

Here, **H** is an n-dimensional unitary space and A, B, and C = A + B are quadratic matrices of order n. In the following, we will not distinguish between a square matrix and the operator acting in **H** defined by it.

THEOREM 2. If the same conditions are imposed on the function $f(x)$ as above, then

$$\mathrm{Sp} f(|C|) \leqslant \mathrm{Sp} f(|A|) + \mathrm{Sp} f(|B|). \tag{2}$$

Two lemmas will be used in the proof of Theorem 2.

LEMMA 1. If A and B are Hermitian matrices and B has only one nonzero eigenvalue, then there exist unitary matrices U and V such that

$$|A+B| \leqslant U^{-1}|A|U + V^{-1}|B|V.$$

PROOF. We can obviously restrict ourselves to the case where the operator B is positive. Let $\{\alpha_i\}$, $\{\beta_i\}$, $\{\gamma_i\}$ be the eigenvalues of the operators A, B, and C = A + B arranged in nonincreasing order and let $\{\varphi_i\}$, $\{\psi_i\}$, $\{\chi_i\}$ be the corresponding eigenvectors. Let α_i, β_i, and γ_i satisfy the relations

$$\beta_1 > 0, \quad \beta_2 = \ldots = \beta_n = 0,$$
$$\alpha_1 \geqslant \alpha_2 \geqslant \ldots \geqslant \alpha_l \geqslant 0 > \alpha_{l+1} \geqslant \ldots \geqslant \alpha_n,$$
$$\gamma_1 \geqslant \gamma_2 \geqslant \ldots \geqslant \gamma_k \geqslant 0 > \gamma_{k+1} \geqslant \ldots \geqslant \gamma_n.$$

It follows from the minimax principle and Weyl's inequality (see [5]) that $l \leq k \leq l+1$ and

$$\alpha_1 + \beta_1 \geqslant \gamma_1 \geqslant \alpha_1 \geqslant \gamma_2 \geqslant \alpha_2 \geqslant \ldots \geqslant \gamma_n \geqslant \alpha_n. \tag{3}$$

Let us consider the operators $|C| = C_+ + C_-$, where

$$C_+ = \sum_{i=1}^{k} \gamma_i (\cdot, \chi_i)\chi_i,$$

$$C_- = -\sum_{i=k+1}^{n} \gamma_i (\cdot, \chi_i)\chi_i,$$

and

$$|A| = A_+ + A_-,$$

where

$$A_+ = \sum_{i=1}^{l} \alpha_i (\cdot, \varphi_i)\varphi_i,$$

$$A_- = -\sum_{i=l+1}^{n} \alpha_i (\cdot, \varphi_i)\varphi_i.$$

Let G denote the k-dimensional space spanned by the vectors $\chi_1, \ldots, \chi_k$ and let Q denote the orthogonal projector onto G. Let us introduce the operators

$$\tilde{C} = QCQ, \quad \tilde{A} = QAQ, \text{ and } \tilde{B} = QbQ.$$

It is clear that $\tilde{C} = C_+$. Let us take

$$\tilde{A} = \sum_{i=1}^{k} \tilde{\alpha}_i (\cdot, \tilde{\varphi}_i)\tilde{\varphi}_i \quad (\tilde{\varphi}_i \in G), \tag{4}$$

$$\tilde{B} = \sum_{i=1}^{k} \tilde{\beta}_i (\cdot, \tilde{\psi}_i)\tilde{\psi}_i \quad (\tilde{\psi}_i \in G). \tag{5}$$

It follows from the minimax principle that

$$\tilde{\alpha}_i \leqslant \alpha_i, \quad i = 1, 2, \ldots, k, \tag{6}$$

$$\tilde{\beta}_1 \leqslant \beta_1, \quad \tilde{\beta}_2 = \ldots = \tilde{\beta}_k = 0. \tag{7}$$

Let us consider unitary operators U, V such that

$$U^{-1}\varphi_i = \begin{cases} \tilde{\varphi}_i & i = 1, \ldots, k \\ \chi_i & i = k+1, \ldots, n, \end{cases}$$
$$V^{-1}\psi_1 = \tilde{\psi}_1.$$

It is not difficult to see that the following inequality follows from (4) and (6):

$$\tilde{A} \leqslant U^{-1}A_+U,$$

whereas (5) and (7) yield

$$\tilde{B} \leqslant V^{-1}BV.$$

Since $C_+ = \tilde{C} = \tilde{A} + \tilde{B}$, the above inequalities yield

$$C_+ \leqslant U^{-1}A_+U + V^{-1}BV,$$

on the basis of (3) we also have

$$C_- \leqslant U^{-1}A_-U,$$

so that by addition we obtain

$$|C| \leqslant U^{-1}|A|U + V^{-1}BV.$$

The lemma has been proved.

LEMMA 2. Let A be a square matrix of order n and let $\hat{A} = \begin{pmatrix} 0 & A^* \\ A & 0 \end{pmatrix}$ be a Hermitian matrix of order 2n. Then, we have

$$s_{2k-1}(\hat{A}) = s_{2k}(\hat{A}) = s_k(A) \quad (k = 1, \ldots, n).$$

This simple but useful lemma has been established by Wielandt (see [4]).

PROOF OF THEOREM 2. In view of Lemma 2, we only have to prove inequality (2) for Hermitian matrices. Next, we can assume that the operator B is one-dimensional. It is now sufficient to prove (2) only for positive operators A and B.

Indeed, let us suppose that inequality (2) is satisfied under these assumptions. Then, if A and B are Hermitian matrices and B is one-dimensional we have

$$|C| \leqslant U^{-1}|A|U + V^{-1}|B|V$$

on account of Lemma 1 and, therefore, we have

$$\operatorname{Sp} f(|C|) \leqslant \operatorname{Sp} f(U^{-1}|A|U + V^{-1}|B|V) \leqslant$$
$$\leqslant \operatorname{Sp} f(U^{-1}|A|U) + \operatorname{Sp} f(V^{-1}|B|V) = \operatorname{Sp} f(|A|) + \operatorname{Sp} f(|B|).$$

Hence, let us take

$$A = \sum_{i=1}^{n} \alpha_i(\cdot, \varphi_i)\varphi_i \quad (\alpha_i \geqslant 0),$$
$$B = \beta(\cdot, \psi)\psi \quad (\beta > 0),$$
$$C = \sum_{i=1}^{n} \gamma_i(\cdot, \chi_i)\chi_i \quad (\gamma_i \geqslant 0).$$

Let $f'(x)$ denote the left derivative of the increasing concave function $f(x)$; as is well known, the derivative exists everywhere and is a nonnegative nonincreasing function. Then, since a concave function is absolutely continuous, we have

$$f(x) = \int_0^x f'(t)\,dt.$$

Next, we have

$$\mathrm{Sp}\, f(C)=\sum_{i=1}^{n} f(\gamma_i)=\sum_{i=1}^{n}\int_0^{\gamma_i} f'(t)\,dt=\sum_{i=1}^{n}\int_0^{\alpha_i} f'(t)\,dt+\sum_{i=1}^{n}\int_{\alpha_i}^{\gamma_i} f'(t)\,dt=\sum_{i=1}^{n} f(\alpha_i)+\sum_{i=1}^{n}\int_{\alpha_i}^{\gamma_i} f'(t)\,dt.$$

Inequality (3) shows that the segments $[\alpha_i, \gamma_i)$ do not overlap. Since $\sum_{i=1}^{n}(\gamma_i-\alpha_i)=\beta$, we have

$$\sum_{i=1}^{n}\int_{\alpha_i}^{\gamma_i} f'(t)\,dt\leqslant\int_0^{\beta} f'(t)\,dt=f(\beta).$$

This completes the proof of Theorem 2.

3. Let us now give the proof of Theorem 1. Here, A and B are completely continuous operators in infinite-dimensional Hilbert space and C = A + B.

Let $C = W|C|$ be the polar representation of the operator C (see [5]) and let

$$|C|=\sum_k s_k(C)(\cdot,\ \chi_k)\,\chi_k.$$

Let M_n denote the subspace spanned by the vectors $\chi_1, \ldots, \chi_n$ and let P_n be the orthogonal-projection operator onto M_n. Theorem 2 can be applied to the operators

$$A_n=P_nW^*AP_n,\quad B_n=P_nW^*BP_n,\quad C_n=P_nW^*CP_n=P_n|C|P_n$$

and according to this theorem we have

$$\sum_{k=1}^{n} f(s_k(C_n))\leqslant\sum_{k=1}^{n} f(s_k(A_n))+\sum_{k=1}^{n} f(s_k(B_n)). \tag{8}$$

Taking into account that

$$\begin{aligned}&s_k(C_n)=s_k(C),\\ &s_k(A_n)\leqslant s_k(A),\\ &s_k(B_n)\leqslant s_k(B)\quad (k=1,\ 2,\ \ldots,\ n)\end{aligned}$$

we arrive at inequality (1). Theorem 1 has been proved.

COROLLARY. Let $F(x)$, where $0 \leq x < +\infty$, be a nondecreasing convex function such that $F(0) = 0$. Then, under the conditions of Theorem 1 we have

$$\sum_{k=1}^{n} F(f(s_k(A+B)))\leqslant\sum_{k=1}^{n} F(f(s_k(A))+f(s_k(B))).$$

The relations can be obtained from inequality (1) by the application of the Karamat–Weyl theorem (see [5]).

Analogs of the Lidskii–Wielandt inequality (see [2, 4]) and the Amir-Moez inequality (see [3]) can be derived from (1) by the methods of [3, 4]. The corresponding results are given below in the form of Theorems 3 and 4.

THEOREM 3. (See [2, 4].) *Under the conditions of Theorem 1, the following inequalities hold for any selection of the indices* $j_1 \leq j_2 \leq \ldots \leq j_n$:

$$\sum_{k=1}^{n} f(s_{j_k}(C))\leqslant\sum_{k=1}^{n} f(s_{j_k}(A))+\sum_{k=1}^{n} f(s_k(B)).$$

THEOREM 4. (See [3].) Let A and B be positive completely continuous operators and let C = A + B. If

$$1 \leqslant i_1 \leqslant i_2 \leqslant \ldots \leqslant i_k \leqslant n, \quad 1 \leqslant j_1 \leqslant j_2 \leqslant \ldots \leqslant j_k \leqslant n$$

is a selection of indices such that $k \le n$ and

$$i_p + j_p \leqslant n - k + p + 1 \quad (p = 1, 2, \ldots, k),$$

then the following inequality holds under the conditions of Theorem 1:

$$\sum_{p=1}^{k} f(\lambda_{m(l_p)}(C)) \leqslant \sum_{p=1}^{k} f(\lambda_{m(i_p)}(A)) + \sum_{p=1}^{k} f(\lambda_{m(j_p)}(B)),$$

where $l_p = i_p + j_p - 1$ $(p = 1, 2, \ldots, k)$ and $m(i_1) = i_1$, $m(i_p) = \max(i_p, m(i_{p-1}) + 1)$ $(p = 2, 3, \ldots, k)$.

We omit the proofs of these theorems as they can be obtained by a word-for-word repetition of the corresponding arguments given in [3, 4].

4. Here and in the following, the functions $f(x)$ and $g(x)$ are assumed to satisfy the conditions of Theorem 1.

Let us consider the class $\mathbf{S}_f$ consisting of all those operators A belonging to $\mathbf{S}_\infty$ for which we have $\sum_k f(s_k(A)) < +\infty$. $\mathbf{S}_f$ is a two-sided ideal in **R**. It transforms into a complete separable linear metric space (in general, nonnormable) if we introduce a metric with the help of the formula

$$\rho_f(A, B) = \sum_k f(s_k(A - B)) \quad (A, B \in S_f).$$

The triangle inequality follows from Theorem 1. The ideal $\mathbf{S}_f$ is always contained in $\mathbf{S}_1$, the ideal of kernel operators (see [5]). Let us note one particular case. If

$$f(x) = \frac{x^p}{1 + \mu x^p} \quad (0 < p \leqslant 1;\ \mu \geqslant 0),$$

then the spaces $\mathbf{S}_p$ $(0 < p \le 1)$ will be the corresponding ideals. In this case, the triangle inequality will be of the form

$$\sum_k \frac{s_k^p(A + B)}{1 + \mu s_k^p(A + B)} \leqslant \sum_k \frac{s_k^p(A)}{1 + \mu s_k^p(A)} + \sum_k \frac{s_k^p(B)}{1 + \mu s_k^p(B)}. \tag{9}$$

A new relation between the Fredholm determinants follows from inequality (9). Indeed, let us assume that $A, B \in \mathbf{S}_p$ $(0 < p \le 1)$. Integrating (9) with respect to μ, we obtain

$$\mathrm{Sp} \ln (I + \mu |A + B|^p) \leqslant \mathrm{Sp} \ln (I + \mu |A|^p) + \mathrm{Sp} \ln (I + \mu |B|^p),$$

which directly leads to

$$\det (I + \mu |A + B|^p) \leqslant \det (I + \mu |A|^p) \cdot \det (I + \mu |B|^p).$$

5. Birman and Solomyak [6, 7] have studied the problem of the continuity of the transformers Φ defined by integrals* of the form

$$\Phi T = \int_\Lambda \int_M \varphi(\lambda, \mu)\, dF_\mu T dE_\lambda$$

* Here, E_λ and F_μ are expansions of unity in **H**. The definitions and properties of integrals of this type have been studied in detail in [6, 7].

in various symmetrically normed ideals of the ring **R**. Here, we will consider the analogous problem for the ideals $\mathbf{S}_f$.

Let $\mathbf{S}_f$ and $\mathbf{S}_g$ be two such ideals. It should be noted that if the bound

$$\rho_f(\Phi T,\ 0) \leqslant C_f g(s), \tag{10}$$

where the constant C_f is independent of ω and θ, holds for the one-dimensional operator

$$T = s(\cdot,\ \omega)\theta \quad (\|\omega\| = \|\theta\| = 1),$$

then the transformer acts continuously from $\mathbf{S}_g$ into $\mathbf{S}_f$. Indeed, for any operator T belonging to $\mathbf{S}_g$,

$$T = \sum_k T_k, \quad T_k = s_k(\cdot,\ \omega_k)\theta_k$$

we have

$$\rho_f(\Phi T,\ 0) \leqslant \sum_k \rho_f(\Phi T_k,\ 0) \leqslant C_f \sum_k g(s_k) = C_f \rho_g(T,\ 0).$$

It has been shown in [6] that the problem of the derivation of inequality (10) reduces to the derivation of bounds in ω and θ on the s-numbers of the integral operator K acting from space $L_{2,\sigma}(\Lambda)$ into the space $L_{2,\tau}(M)$ according to the formula

$$Ku(\lambda) = \int_\Lambda \varphi(\lambda,\ \mu)\, u(\lambda)\, d\sigma(\lambda),$$

where

$$\sigma(\lambda) = (E_\lambda\omega,\ \omega), \quad \tau(\mu) = (F_\mu\theta,\ \theta).$$

With the help of the bounds on the s-number of integral operators given in [8], we can easily formulate the conditions under which the transformer Φ in the spaces $\mathbf{S}_f$ is continuous. For example, let us give the conditions for the continuity of Φ in $S_p (0 < p \leq 1)$.

THEOREM 5. Let $\Lambda = M = Q^m$ be a unit cube in Euclidean space Em. If $\varphi(\cdot,\ \mu)$ belongs to $W_q^\alpha(Q^m)$ for almost all μ ($q\alpha > m$; $q < 2$) and

$$\tau\text{-sup}\, \|\varphi(\cdot,\ \mu)\|_{W_q^\alpha} \leqslant L,$$

then the transformer Φ acts continuously from S_p into S_p with

$$p > \left(\frac{\alpha}{m} + 1 - \frac{1}{q}\right)^{-1}.$$

In conclusion, I would like to express my gratitude to M. Sh. Birman for suggesting the problem to me and for his supervision of the work.

LITERATURE CITED

1. Ky Fan, Proc. Nat. Acad. Sci., USA, Vol. 36, No. 1 (1950).
2. V. B. Lidskii, Dokl. Akad. Nauk SSSR, Vol. 75, No. 6 (1950).
3. A. Amir-Moez, Duke Math. J., Vol. 23, No. 3 (1956).
4. A. S. Markus, Uspekhi Mat. Nauk, Vol. 19, No. 4 (1964).
5. I. Ts. Gokhberg and M. G. Krein, Introduction to the Theory of Linear Nonself-adjoint Operators [in Russian], Izd. Nauka (1966).
6. M. Sh. Birman and M. Z. Solomyak, "Stieltjes double-integral operators," in: Topics in Mathematical Physics, Vol. 1, Consultants Bureau, New York (1967).
7. M. Sh. Birman and M. Z. Solomyak, "Stieltjes double-integral operators. II," in Topics in Mathematical Physics, Vol. 2, Consultants Bureau, New York (1968).
8. M. Sh. Birman and M. Z. Solomyak, Vestnik LGU, No. 7 (1967).
9. S. Yu. Rotfel'd, Funktsional. Analiz i Ego Prilozhen., Vol. 1, No. 3 (1967).

DOUBLE-INTEGRAL OPERATORS IN THE RING $\hat{R}$

M. Z. Solomyak

1. This article is devoted to a study of double-integral operators of the form

$$\int_{\Lambda}\int_{M} f(\lambda, \mu) F(d\mu) E(d\lambda). \tag{1}$$

Here, $E(\cdot)$, $F(\cdot)$ are spectral measures in separable Hilbert space **H**, Λ and M are their carriers, and f a scalar function on $\Lambda \times M$. Integrals (1) represent a particular case of the integrals

$$\int_{\Lambda}\int_{M} f(\lambda, \mu) F(d\mu)\, TE(d\lambda)$$

(T is a bounded operator in **H**) which have been studied in detail by Birman and Solomyak [1-4]. The results obtained in these articles lead, in particular, to some conditions for the existence of the integral (1) formulated in terms of the smoothness of function f; the necessary degree of smoothness increases with the number of dimensions of space Λ (or M). It should be noted that in this context Λ is assumed to be a smooth finite-dimensional manifold and $E(\cdot)$ is assumed to be a Borel measure.

By contrast with [1-4], the integrals (1) in the present paper are considered to within an arbitrary completely continuous additive term. More accurately, we do not study the individual operators of the form (1), but the residue class of the ring of all bounded operators in **H** modulo the ideal of all completely continuous operators. It is found that under certain conditions on the spectral measures $E(\cdot)$ and $F(\cdot)$, and integral of the form (1) can be reasonably defined (as an element of the residue ring) for any continuous function f.

The procedure presented in the present article provides a good model of the situation that is well known in the theory of multidimensional singular integrals. Namely, a singular integral operator Ω with symbol $\omega(\theta, x)$ (where $\theta \in S^{m-1}$, $x \in R^m$), considered as an operator in $L_2(R^m)$, is a bounded operator if the symbol is a sufficiently smooth function (see [5]-[7]). On the other hand, it has recently been found (see [8]-[13]) that

$$\inf \|\Omega - T\| = \max |\omega|, \tag{2}$$

where the exact lower limit is taken over the set of all completely continuous operators T. Expression (2) allows us to associate a residue class with any continuous symbol ω (ω has to be continuous everywhere, including the point at infinity). It is natural to call any operator belonging to this class a singular integral operator with symbol ω.

The analogy between integrals (1) and singular integrals is not a chance one: It has been shown in [3], [4] that a singular operator can be written in the form of (1), where $E(\cdot)$, $F(\cdot)$ are now specially defined spectral measures and the operator symbol plays the part of the function f. This approach has been used in [3], [4], and [14] to obtain many well-known results, as well as new criteria for the boundedness of singular operators in $L_2(R^m)$. The results of the present article show that the theory of singular integrals as elements of the residue ring can also be developed on the basis of the theory of double-integral operators.

2. This section contains the necessary background taken from the theory of operators. We use **R** to denote the ring of all bounded operators in **H** and $\mathbf{S}_\infty$ to denote the ideal of all completely continuous operators of **R**. It is customary to denote the quotient-ring $\mathbf{R}/\mathbf{S}_\infty$ by the symbol $\hat{\mathbf{R}}$. The elements of the ring $\hat{\mathbf{R}}$ are classes of the operators of **R**, two operators belonging to the same class if and only if their difference is completely continuous. Operations and involutions in $\hat{\mathbf{R}}$ are induced by corresponding operations in **R**. The ring $\hat{\mathbf{R}}$ represents a Banach algebra with respect to the norm

$$\|\mathbf{A}\|_{\hat{R}} = \inf_{A \in \mathbf{A}} \|A\|_R.$$

The class which contains a specific operator A belonging to **R** is denoted by $\hat{A}$. The norm of class $\hat{A}$ is called the essential norm ($\|A\|_{ess}$) of operator A. In other words, we have

$$\|A\|_{ess} = \|\hat{A}\|_{\hat{R}} = \inf_{T \in S_\infty} \|A - T\|_R.$$

The concept of essential norm is closely connected with the concept of essential spectrum (condensation spectrum) of an operator. Let us consider the set **X** of all possible sequences $\{x_n\}$ of elements of **H** such that $\|x\| = 1$ $(n = 1, 2, \dots)$ and $x_n \to 0$. We say that a complex number z belongs to the essential spectrum of an operator A $(z \in \sigma_{ess}(A))$, if we can find a sequence $\{x_n\}$ belonging to **X** such that

$$\|Ax_n - zx_n\| \to 0.$$

The closed set $\sigma_{ess}(A)$ is independent of the choice of the operator A belonging to **A** and coincides with the spectrum of the element $\hat{\mathbf{A}}$ of ring $\hat{\mathbf{R}}$.

For any operator A belonging to **R** we have

$$\|A\|_{ess} \geq \max_{z \in \sigma_{ess}(A)} |z|. \tag{3}$$

If A is a symmetric operator or differs from a symmetric one by completely continuous operator (in other words, if $\hat{A} = \hat{A}^*$), then

$$\|A\|_{ess} = \max_{z \in \sigma_{ess}(A)} |z|. \tag{4}$$

In the case of nonself-adjoint operators, we have (see [15], Chapter II, Section 7)

$$\|A\|_{ess}^2 = \|A^*A\|_{ess}, \tag{5}$$

so that the calculation of the essential norm reduces to the case of a symmetric operator. Equation (5) means that the $\hat{\mathbf{R}}$ algebra is a B* algebra [16].

In many cases it is more convenient to use another formula for the essential norm. In fact, for any operator A belonging to **R** we have

$$\|A\|_{ess} = \sup_{\{x_n\} \in X} \overline{\lim_n} \|Ax_n\|, \tag{6}$$

which can be easily derived from Equation (5). The following lemma which will be used in the following can be simply established with the help of formula (6).

LEMMA 1. Let $A_2 > A_1 > 0$, $B_2 > B_1 > 0$ be two pairs of operators belonging to **R** such that A_1 commutes with A_2 and B_1 commutes with B_2. Then, we have

$$\|A_1B_1\|_{ess} \leqslant \|A_2B_2\|_{ess}.$$

PROOF. It is obviously sufficient to consider the case $B_1 = B_2 = B$. In this case, using formula (6) and the inequality $\|A_1Bx\| \leq \|A_2Bx\|$ which is valid for any $x \in \mathbf{H}$, we obtain

$$\|A_1B\|_{ess} \leqslant \|A_2B\|_{ess}.$$

The lemma has been proved.

3. Let Λ, M be compact metric spaces, one of which (M for definiteness) has a finite number of dimensions.* Let $C(\Lambda)$, $C(M)$, and $C(\Lambda \times M)$ denote, as is usual, the spaces of the continuous functions on the corresponding compactums. Let $E(\cdot)$, $F(\cdot)$ be the spectral measures defined on the Borel subspaces of Λ and M, respectively. If we have $\varphi \in C(\Lambda)$ and $\chi \in C(M)$, then by convention we will write

$$\Phi = \int\limits_{\Lambda} \varphi(\lambda)\, E(d\lambda), \quad \Psi = \int\limits_{M} \psi(\mu)\, F(d\mu).$$

All subsequent results depend on the following assumption concerning the spectral measures $E(\cdot)$, $F(\cdot)$:

PRINCIPAL ASSUMPTION. For any functions φ and χ, where $\varphi \in C(\Lambda)$, $\chi \in C(M)$, the commutator of the operators Φ and Ψ is completely continuous, i.e.,

$$\Phi\Psi - \Psi\Phi \in S_\infty \quad (\varphi \in C(\Lambda),\ \psi \in C(M)). \tag{7}$$

With the help of this assumption we can give a strict rigorous meaning to the integral

$$Q_f = \int\limits_{\Lambda}\int\limits_{M} f(\lambda, \mu)\, \hat{F}(d\mu)\, \hat{E}(d\lambda) \quad (f \in C(\Lambda \times M)); \tag{8}$$

we can show that the set $\hat{Q} = \{Q_f\colon f \in C(\Lambda \times M)\}$ represents a commutative subring of $\hat{R}$; we can establish that $\hat{Q}$ is isometrically isomorphic to the ring $C(K)$, where $K \subset \Lambda \times M$ is some compactum (which thus realizes the bicompactum of the maximal ideals of ring $\hat{Q}$); finally, we can provide a description of set K in terms of the spectral measures $E(\cdot)$ and $F(\cdot)$.

The facts listed above allow us to give a simple interpretation in the language of the theory of commutative normed rings. Let us consider the rings of the operators $\mathbf{C}(\Lambda) = \{\Phi\colon \varphi \in C(\Lambda)\}$ and $\mathbf{C}(M) = \{\Psi\colon \psi \in C(M)\}$. These rings do not commute with each other; however, condition (7) means that the rings

$$\hat{C}(\Lambda) = \{\hat{\Phi}\colon \varphi \in C(\Lambda)\} \quad \text{and} \quad \hat{C}(M) = \{\hat{\Psi}\colon \psi \in C(M)\}$$

do commute. The ring $\hat{Q}$ represents nothing else but a commutative subalgebra of the $\hat{R}$ algebra generated by $\hat{C}(\Lambda)$ and $\hat{C}(M)$. It follows from this that the space of the maximal ideals of ring $\hat{Q}$ is realized as a compact subset $K \subset \Lambda \times M$. Since $\hat{Q}$, being a subalgebra of $\hat{R}$, is itself a B* algebra, the ring $\hat{Q}$ according to a theorem of Gel'fand and Naimark [16] is isometrically isomorphic to the ring of function continuous on K.

* Some remarks will be made in Section 5. It should be noted that the assumptions concerning the spaces Λ, M are not always used to the same extent. In particular, Λ, M can be considered in Section 4 to be arbitrary bicompact normal topological spaces.

Such a "ring" viewpoint has been used in the theory of singular integrals beginning with Gokhberg [8]; the most successful application can be found in Cordes' article [13].

The theory of normed rings is not used in the derivations presented below. It should be noted in this connection that there are two aspects of the work which cannot be treated within the framework of ring theory and which, possibly, are of independent interest. The first of these is the realization of elements of ring $\hat{Q}$ as limits in the norm $\|\cdot\|_R$ of "integral" sums; the second is the description of the set of maximal ideals of this ring in terms of the spectral measures $E(\cdot)$ and $F(\cdot)$.

Our approach has the advantage that it makes possible the extension of some of the results of this article to the case of discontinuous functions f where the theory of commutative normed rings is completely inapplicable. In the present article, however, we restrict ourselves to the study of integrals of continuous functions.

4. It is convenient for us to start with the construction of the set K and the investigation of its structure. The investigation is based on the following lemma.

LEMMA 2. *Let* $\delta_1 \subset \delta_2 \subset \Lambda$, $\partial_1 \subset \partial_2 \subset M$ *be Borel sets. Let us assume that there exist real functions* $\varphi \in C(\Lambda)$, $\psi \in C(M)$ *such that**

$$\chi_{\delta_1}(\lambda) \leqslant \varphi(\lambda) \leqslant \chi_{\delta_2}(\lambda), \qquad \lambda \in \Lambda, \tag{9}$$

$$\chi_{\partial_1}(\mu) \leqslant \psi(\mu) \leqslant \chi_{\partial_2}(\mu), \qquad \mu \in M. \tag{10}$$

Then, we have either

$$F(\partial_1)E(\delta_1) \in \mathbf{S}_\infty, \tag{11}$$

or

$$\|F(\partial_2)E(\delta_2)\|_{ess} = 1. \tag{12}$$

PROOF. Let us assume that (11) does not hold. From formula (6) we can now find a number $\gamma > 0$ and a sequence $\{x_n\} \in \mathbf{X}$ such that

$$\|F(\partial_1)E(\delta_1)x_n\| \geqslant \gamma, \quad n = 1, 2, \ldots \tag{13}$$

We can obviously assume that

$$E(\delta_1)x_n = x_n, \ n = 1, 2 \ldots$$

Let us consider the sequence $\{y_n\} = \{\Psi x_n\}$. From inequalities (10) and (13) we find that

$$\|F(\partial_2)y_n\| = \|y_n\| \geqslant \|F(\partial_1)x_n\| \geqslant \gamma.$$

Let us next use inequalities (9) and (10) and condition (7),

$$\|y_n - E(\delta_2)y_n\| \leqslant \|y_n - \Phi y_n\| = \|(\Psi\Phi - \Phi\Psi)x_n\| \to 0.$$

The last two relations show that for the sequence $\{\tilde{y}_n\} = \{\|y\|^{-1}y_n\}$ we have

$$\|E(\delta_2)F(\partial_2)\tilde{y}_n\| \to 1.$$

Since we obviously have $\{\tilde{y}_n\} \in \mathbf{X}$, this together with (6) yields

$$\|E(\delta_2)F(\partial_2)\|_{ess} = \|F(\partial_2)E(\delta_2)\|_{ess} = 1.$$

The lemma has been proved.

The conditions of Lemma 2 are satisfied, for example, if the sets δ_1, ∂_1 are closed and δ_2, ∂_2 are open. If, in particular, the set δ is simultaneously open and closed in Λ, then we can set $\delta_1 = \delta_2 = \delta$.

* As is usual, χ_e denotes the characteristic function of the set e.

Let us now consider the set $K \subset \Lambda \times M$. We will say that the point (λ_0, μ_0), where $(\lambda_0, \mu_0) \in \Lambda \times M$, belongs to K if and only if the following relation holds for any neighborhoods $\delta \ni \lambda_0$, $\partial \ni \mu_0$

$$\| F(\partial) E(\delta) \|_{ess} = 1.$$

It is easy to see that K is closed; it follows from Lemma 2 that if $(\lambda_0, \mu_0) \bar{\in} K$, then we can find neighborhoods $\delta \ni \lambda_0$, $\partial \ni \mu_0$ such that $F(\partial) E(\delta) \in S_\infty$.

Let us study the structure of K in greater detail. Let Λ_0 denote the carrier of the essential part of the spectral measure $E(\cdot)$: the point $\lambda_0 \in \Lambda$ belongs to Λ_0 if and only if the projector $E(\delta)$ is infinite-dimensional for any neighborhood $\delta \ni \lambda_0$. We introduce the set $M_0 \subset M$ in an analogous manner. Let π_1 and π_2 denote the canonical mappings of the product $\Lambda \times M$ onto Λ and M, i.e., $\pi_1[(\lambda, \mu)] = \lambda$ and $\pi_2[(\lambda, \mu)] = \mu$.

LEMMA 3. We have

$$\pi_1 K = \Lambda_0; \quad \pi_2 K = M_0. \tag{14}$$

PROOF. The proof will only be given for the first of the relations (14). The inclusion $\pi_1 K \subset \Lambda_0$ is obvious; let us establish the inverse inclusion. Let $\lambda_0 \in \Lambda_0 \setminus \pi_1 K$. Then for any point $\mu \in M$ we can find neighborhoods $\delta_\mu \ni \lambda_0$, $\partial_\mu \ni \mu$ such that $\| F(\partial_\mu) E(\delta_\mu) \|_{ess} = 0$. Let us choose from the covering of the compactum M by open sets ∂_μ a finite covering $\{\partial_{\mu_j}\}$, $j = 1, \ldots, j_0$, and let us consider the neighborhood of the point λ_0

$$\delta = \bigcap_{j=1}^{j_0} \delta_{\mu_j}.$$

Since we have $\sum_{j=1}^{j_0} F(\partial_{\mu_j}) \geqslant I$, by Lemma 1 we obtain

$$1 = \| E(\delta) \|_{ess} \leqslant \left\| \sum_{j=1}^{j_0} F(\partial_{\mu_j}) E(\delta) \right\|_{ess} \leqslant \sum_{j=1}^{j_0} \| F(\partial_{\mu_j}) E(\delta) \|_{ess} = 0.$$

This contradiction shows us that $\pi_1 K \supset \Lambda_0$. The lemma has been proved.

Let us now consider two particular cases. Let $\Lambda = M$, $E(\delta) = F(\delta)$. Then, it is obvious that the set K represents the diagonal $\lambda = \mu$ $(\lambda \in \Lambda_0)$. On the other hand, if $E(\cdot)$, $F(\cdot)$ are the spectral measures that generate the set of singular integral operators in $L_2(R^m)$ (see [3], [4]), then $K = \Lambda \times M = S^m \times S^{m-1}$. This follows from the well-known lemma of Gokhberg [8, 9].

5. Before we proceed to the construction of the integral (8), let us make several remarks concerning the covering of compact spaces.

Let $\overset{\circ}{\partial} = \{\partial_j\}$, $j = 1, \ldots, j_0$, be a finite covering of the compact metric space M by open sets. With each such covering we can associate (nonuniquely) a collection of nonnegative functions $\overset{\circ}{\psi} = \{\psi_j\}$, $j = 1, \ldots, j_0$, which are continuous on M and which satisfy the conditions

$$\sum_{j=1}^{j_0} \psi_j(\mu) \equiv 1 \qquad (\mu \in M);$$

$$\psi_j(\mu) = 0 \; (\mu \notin \partial_j), \qquad j = 1, \ldots j_0.$$

Any selection of functions $\mathring{\psi}$ satisfying these conditions will be called the decomposition of unity on M coordinated with the covering $\mathring{\partial}$.

The quantity $d(\mathring{\partial}) = \max_j \operatorname{diam} \partial_j$ is called the diameter of the covering $\mathring{\partial}$. Any covering for which $d(\mathring{\partial}) < \varepsilon$ is called a ε-covering.

Let us determine for any $\alpha = 1, \ldots, j_0$ the number r_α which is equal to the number of sets $\partial_j \in \mathring{\partial}$ which have a nonnull intersection with ∂_α. The number $r(\mathring{\partial}) = \max_\alpha r_\alpha$ will be called the rank of the covering $\mathring{\partial}$. If the number of dimensions of M is finite, then, as can be easily shown, there exists a number r(M) such that for any $\varepsilon > 0$ we can find an ε-covering whose rank does not exceed r(M).

Let $\mathring{\partial}' = \{\partial'_j\}$, $j = 1, \ldots, j_0$, and $\mathring{\partial}'' = \{\partial''_l\}$, $l = 1, \ldots, l_0$, be two coverings of M. The covering $\mathring{\partial} = \{\partial_{jl}\} = \{\partial'_j \cap \partial''_l\}$ will be called the product of the coverings $\mathring{\partial}'$ and $\mathring{\partial}''$ and will be denoted by $\mathring{\partial}' \circ \mathring{\partial}''$. We do not exclude the case when $\mathring{\partial}' = \mathring{\partial}''$. It is obvious that we have $d(\mathring{\partial}' \circ \mathring{\partial}'') \leqslant \min(d(\mathring{\partial}'), d(\mathring{\partial}''))$ and $r(\mathring{\partial}' \circ \mathring{\partial}'') \leqslant r(\mathring{\partial}') \, r(\mathring{\partial}'')$. If $\mathring{\psi}'$ and $\mathring{\psi}''$ are decompositions of unity coordinated with the coverings $\mathring{\partial}'$ and $\mathring{\partial}''$, respectively, then the collection $\mathring{\psi}' \circ \mathring{\psi}'' = \{\psi_{jl}\} = \{\psi'_j \psi''_l\}$ forms a decomposition of unity coordinated with the covering $\mathring{\partial}' \circ \mathring{\partial}''$.

6. Here we consider elements of the ring $\hat{R}$ which have a special form and which in the following play the part of integral sums for the integral (8). Let $\mathring{\delta} = \{\delta_i\}$, $i = 1, \ldots, i_0$, be an open covering of space Λ and $\mathring{\varphi} = \{\varphi_i\}$ be a decomposition of unity on Λ coordinated with it. Analogously, $\mathring{\partial} = \{\partial_j\}$, $j = 1, \ldots, j_0$, is an open covering of space M and $\psi = \{\psi_j\}$ is the decomposition of unity on M coordinated with it. Let $\mathring{c} = \{c_{ij}\}$, $i = 1, \ldots, i_0$, $j = 1, \ldots, j_0$, be a selection of complex numbers. Let us consider the operator

$$A = A(\mathring{\varphi}, \mathring{\psi}, \mathring{c}) = \sum_{i=1}^{i_0} \sum_{j=1}^{j_0} c_{ij} \Psi_j \Phi_i \tag{15}$$

and the residue class

$$\hat{A} = \hat{A}(\mathring{\varphi}, \mathring{\psi}, \mathring{c}) = \sum_{i=1}^{i_0} \sum_{j=1}^{j_0} c_{ij} \hat{\Psi}_j \hat{\Phi}_i \tag{16}$$

defined by it. First of all, in view of condition (7) it is obvious that the classes $\hat{A}$ corresponding to different systems $\mathring{\varphi}$, $\mathring{\psi}$, and $\mathring{c}$ commute with one another. It follows from the definition of K that the summation in (16) can only extend over the pairs of indices (i, j) for which the closure of $\delta_i \times \partial_j$ has a nonnull intersection with K. Let us denote this set by $J = J(\mathring{\delta}, \mathring{\partial})$,

$$J = J(\mathring{\delta}, \mathring{\partial}) = \{(i, j) : (\bar{\delta}_i \times \bar{\partial}_j) \cap K \neq \varnothing\}.$$

Indeed, the other pairs of indices correspond to zero terms in the sum (16)

Our principal problem is the estimation of the norm of the element (16) [or, what is the same, the estimation of the essential norm of operator (15)] as a function of the coefficients $\mathring{c}$. Let us write

$$M = M(\mathring{c}) = \max_{i,j} |c_{ij}|, \qquad m = m(\mathring{\delta}, \mathring{\partial}, \mathring{c}) = \max_{(i,j) \in J} |c_{ij}|.$$

LEMMA 4. We have the bound

$$\|\hat{A}(\mathring{\varphi}, \mathring{\psi}, \mathring{c})\|_{\hat{R}} \leqslant r(\mathring{\partial})\, m(\mathring{\delta}, \mathring{\partial}, \mathring{c}). \tag{17}$$

If the coefficients $\mathring{c}$ are such that the inequality

$$|c_{ij_1} - c_{ij_2}| \leqslant \eta, \; i = 1, \ldots, i_0$$

holds for $\partial_{j_1} \cap \partial_{j_2} \neq \varnothing$, then the bound

$$\|\hat{A}(\mathring{\varphi}, \mathring{\psi}, \mathring{c})\|_{\hat{R}} \leqslant M(c) + \eta r^2(\mathring{\partial}) \tag{18}$$

holds in addition to (17).

PROOF. Let us first of all assume that we have $\hat{A} = \hat{A}^*$, i.e., that the operator (15) is symmetric or differs from a symmetric operator by a completely continuous operator. In this case, we can use (4) to estimate the essential norm. Thus, let us assume that $z \in \sigma_{ess}(A)$; then, we can find a sequence $\{x_n\} \in \mathbf{X}$ such that

$$\|Ax_n - zx_n\| = \left\| \sum_{i=1}^{i_0} \sum_{j=1}^{j_0} c_{ij} \Psi_j \Phi_i x_n - zx_n \right\| \to 0. \tag{19}$$

The operators Ψ_j satisfy the relation $\sum_{j=1}^{j_0} \Psi_j = I$. Because of this, we can always find among the sequences $\{\Psi_j x_n\}_{n=1}^{\infty}$, $j = 1, \ldots, j_0$, weakly converging to zero at least one sequence $\{y_n\} = \{\Psi_1 x_n\}$. Changing the enumeration if necessary, we can assume that $\psi_1 \psi_j \equiv 0$ for $j > j_1$, where $j_1 \leq r(\mathring{\delta})$. Let us introduce the operator

$$A' = \sum_{j=1}^{j_1} \sum_{i:(i,j) \in J} c_{ij} \Psi_j \Phi_i.$$

Since we have

$$\Psi_1[Ax_n - zx_n] = \sum_{j=1}^{j_1} \sum_{i=1}^{i_0} c_{ij} \Psi_j [\Psi_1 \Phi_i - \Phi_i \Psi_1] x_n + \sum_{j=1}^{j_1} \sum_{i:(i,j) \in J} c_{ij} \Psi_j \Phi_i y_n + [A' y_n - z y_n],$$

in view of condition (7), the definition of the set $J(\mathring{\delta}, \mathring{\partial})$, and relation (19), we obtain

$$\|A' y_n - z y_n\| \to 0.$$

It follows from this that the number z belongs to the essential spectrum of the operator A' and, consequently, we have

$$|z| \leqslant \|A'\| = \left\| \sum_{j=1}^{j_1} \Psi_j \sum_{i:(i,j) \in J} c_{ij} \Phi_i \right\|.$$

Since the norm of each of the operators $\sum_{i:(i,j) \in J} c_{ij} \Phi_i$ does not exceed m and $\|\Psi_j\| \leq 1 (j = 1, \ldots, j_0)$, we have

$$|z| \leqslant j_1 m \leqslant mr(\partial).$$

This, together with (4) leads to inequality (17).

The number z obviously also belongs to the essential spectrum of the operator

$$\widetilde{A}=\sum_{j=1}^{j_1}\Psi_j\sum_{i=1}^{i_0}c_{ij}\Phi_i=\sum_{j=1}^{j_1}\Psi_j\sum_{i=1}^{i_0}c_{i1}\Phi_i+\sum_{j=1}^{j_1}\Psi_j\sum_{i=1}^{i_0}(c_{ij}-c_{i1})\Phi_i\equiv\widetilde{A}_1+\widetilde{A}_2,$$

which differs from A' by a completely continuous term. This yields

$$|z|\leqslant\|\widetilde{A}_1\|+\|\widetilde{A}_2\|.$$

The norm of $\widetilde{A}_2$, estimated in the same way as above, is

$$\|\widetilde{A}_2\|\leqslant\eta r(\mathring{\partial}).$$

For operator $\widetilde{A}_1$ we have

$$\|\widetilde{A}_1\|\leqslant\left\|\sum_{j=1}^{j_1}\Psi_j\right\|\cdot\left\|\sum_{i=1}^{i_0}c_{i1}\Phi_i\right\|\leqslant M.$$

Combining these inequalities and applying (4), we obtain

$$\|A(\mathring{\varphi},\ \mathring{\psi},\ \mathring{c})\|_{\mathrm{ess}}\leqslant M(\mathring{c})+\eta r(\mathring{\partial}). \tag{20}$$

Let us make use of equality (5) to obtain the estimates (17) and (18) for the general case $\hat{A}\neq\hat{A}^*$. Let A be an arbitrary operator of the form (15). Then, in view of assertion (7), we have

$$\hat{A}^*\hat{A}=\sum_{i=1}^{i_0}\sum_{k=1}^{i_0}\sum_{j=1}^{j_0}\sum_{l=1}^{j_0}c_{ij}\overline{c_{kl}}\hat{\Psi}_j\hat{\Psi}_l\hat{\Phi}_i\hat{\Phi}_k=\hat{A}(\mathring{\varphi}\circ\mathring{\varphi},\ \mathring{\psi}\circ\mathring{\psi},\ \mathring{\gamma}),$$

where $\mathring{\gamma}=\{\gamma_{ik,jl}\}=\{c_{ij}\overline{c_{kl}}\}$ $(i,\ k=1,\ \ldots,\ i_0;\ \ j,\ l=1,\ \ldots,\ j_0)$. Since this is a self-conjugate residue class, we can use relations of the form of (17) and (20) to obtain an estimate of its norm.

It is clear that if $(ik,\ jl)\in I(\mathring{\delta}\circ\mathring{\delta},\ \mathring{\partial}\circ\mathring{\partial})$, then $(i,\ j)\in I(\mathring{\delta},\ \mathring{\partial})$ and $(k,\ l)\in I(\mathring{\delta},\ \mathring{\partial})$. Consequently, we have

$$m(\mathring{\delta}\circ\mathring{\delta},\ \mathring{\partial}\circ\mathring{\partial},\ \mathring{\gamma})\leqslant[m(\mathring{\delta},\ \mathring{\partial},\ \mathring{c})]^2.$$

Since we have $r(\mathring{\partial}\circ\mathring{\partial})\leq r^2(\mathring{\partial})$, this yields

$$\|\hat{A}(\mathring{\varphi},\ \mathring{\psi},\ \mathring{c})\|^2_{\hat{R}}=\|\hat{A}(\mathring{\varphi}\circ\mathring{\varphi},\ \mathring{\psi}\circ\mathring{\psi},\ \mathring{\gamma})\|_{\hat{R}}\leqslant r^2(\mathring{\partial})m^2(\mathring{\delta},\ \mathring{\partial},\ \mathring{c}).$$

This proves estimate (17). Next, if the sets $\partial_{j_1l_1}$ and $\partial_{j_2l_2}$ have a nonnull intersection, then $\partial_{j_1}\cap\partial_{j_2}\neq\varnothing$ and $\partial_{l_1}\cap\partial_{l_2}\neq\varnothing$. On the basis of this it is easy to conclude that if we have $\partial_{j_1l_1}\cap\partial_{j_2l_2}\neq\varnothing$, then the inequality

$$|\gamma_{ik,\ j_1l_1}-\gamma_{ik,\ j_2l_2}|\leqslant 2M(\mathring{c})\eta,\quad i,\ k=1,\ \ldots,\ i_0$$

is satisfied. With the help of inequality (20) this now leads to

$$\|\hat{A}(\mathring{\varphi},\ \mathring{\psi},\ \mathring{c})\|^2_{\hat{R}}=\|\hat{A}(\mathring{\varphi}\circ\mathring{\varphi},\ \mathring{\psi}\circ\mathring{\psi},\ \mathring{\gamma})\|_{\hat{R}}\leqslant M^2(\mathring{c})+2M(\mathring{c})\eta r^2(\mathring{\partial}).$$

It is obvious that inequality (18) now follows. The lemma has been proved.

The following lemma will be required for the derivation of the lower bound on the norm of an element of $\mathbf{Q}_f$.

LEMMA 5. Let $A = A(\mathring{\varphi}, \mathring{\psi}, \mathring{c})$ be an operator of type (15), let (λ_0, μ_0) be a point belonging to set K, and let $c_{ji} = c_0$ for all numbers i, j such that $(\lambda_0, \mu_0) \in \bar{\delta} \times \bar{\partial}_j$. Then, we have

$$\|A\|_{ess} \geqslant |c_0|.$$

PROOF. According to (3) it is sufficient to show that

$$c_0 \in \sigma_{ess}(A).$$

Let $\delta_0 \subset \Lambda$ $(\partial_0 \subset M)$ be a neighborhood of the point λ_0 (or μ_0) such that $\delta_0 \cap \delta_i = \varnothing$ holds if λ_0 $\bar{\in}$ $\bar{\delta}_i$ (or ∂_i $(\partial_0 \cap \partial_j = \varnothing$ holds if μ_0 $\bar{\in}$ $\bar{\partial}_j$). Since we have $\|F(\partial_0)E(\delta_0)\|_{ess} = 1$ onthe account ofthe inclusion $(\lambda_0, \mu_0) \in K$, we can find a sequence $\{x_n\} \in \mathbf{X}$ such that $E(\delta_0)x_n = x_n$ (where n = 1, 2, . . .) and $\|F(\partial_0)x_n\| \to 1$. It is easy to see that the following relations hold:

$$\Psi_j \Phi_i x_n \to 0,$$

provided that (λ_0, μ_0) $\bar{\in}$ $\bar{\delta}_i \times \bar{\partial}_j$;

$$\sum_{i:\lambda_0 \in \bar{\delta}_i} \Phi_i x_n = x_n, \quad n = 1, 2, \ldots,$$

$$\sum_{j:\mu_0 \in \bar{\partial}_j} \Psi_j x_n - x_n \to 0.$$

Using these relations, we find that

$$\lim_n \left\| \sum_{i=1}^{i_0} \sum_{j=1}^{j_0} c_{ij} \Psi_j \Phi_i x_n - c_0 x_n \right\| = \lim_n \left\| \sum_{(i,j):(\lambda_0,\mu_0) \in \bar{\delta}_i \times \bar{\partial}_j} c_0 \Psi_j \Phi_i x_n - c_0 x_n \right\| =$$
$$= |c_0| \lim_n \left\| \sum_{j:\mu_0 \in \bar{\partial}_j} \Psi_j \sum_{i:\lambda_0 \in \bar{\delta}_i} \Phi_i x_n - x_n \right\| = 0,$$

which was to be proved.

7. Let us now proceed to the construction of integral (8). Let us take a fixed value of the number $r \geq r(M)$. Let $\mathring{\delta} = \{\delta_i\}$, $i = 1, \ldots, i_0$, and $\mathring{\partial} = \{\partial_j\}$, $j = 1, \ldots, j_0$, be any open coverings of spaces Λ and M, respectively, and let $r(\mathring{\partial}) \leq r$; let $\mathring{\varphi}$ and $\mathring{\psi}$ be the decompositions of unity on Λ and M coordinated with these coverings. Let us choose an arbitrary point $\xi_{ij} = (\lambda_{ij}, \mu_{ij})$ belonging to $\delta_i \times \partial_j$ $(i = 1, \ldots, i_0;\ j = 1, \ldots, j_0)$; let us denote the collection of these points by $\xi(\mathring{\delta}, \mathring{\partial})$. If f is a function defined on $\Lambda \times M$, then we will denote the set of numbers $\{f(\xi_{ij})\}$ by $f[\mathring{\xi}(\mathring{\delta}, \mathring{\partial})]$. Let us now consider the "integral sum"

$$\hat{A}(\mathring{\varphi}, \mathring{\psi}, f[\mathring{\xi}(\mathring{\delta}, \mathring{\partial})]). \tag{21}$$

We will say that the element Q_f of $\hat{\mathbf{R}}$ is the limit of the integral sums (21) if for every number $\vartheta > 0$ we can find an $\varepsilon > 0$ such that for any ε-coverings $\mathring{\delta}$ and $\mathring{\partial}$ [with $r(\mathring{\partial}) \leq r$], any decompositions of unity $\mathring{\varphi}$, $\mathring{\psi}$ coordinated with them, and for an arbitrary choice of the points $\mathring{\xi}(\mathring{\delta}, \mathring{\partial})$ we have

$$\|\hat{A}(\mathring{\varphi}, \mathring{\psi}, f[\mathring{\xi}(\mathring{\delta}, \mathring{\partial})]) - Q_f\|_{\hat{R}} < \vartheta.$$

If this limit exists, then we will say that it is the integral of the function f over the set functions $\hat{F}(\cdot)\hat{E}(\cdot)$ in ring $\hat{R}$ and we will write

$$Q_f = \int_\Lambda \int_M f(\lambda, \mu)\hat{F}(d\mu)\hat{E}(d\lambda). \tag{8'}$$

The main result of the present paper can be formulated as the following theorem.

THEOREM. If we have $f \in C(\Lambda \times M)$, then integral (8') exists. The set $\hat{Q} = \{Q_f : f \in C(\Lambda \times M)\}$ represents a commutative subring of ring $\hat{R}$ isometrically isomorphic to ring $C(K)$.

PROOF. Let us consider the two integral sums

$$\begin{aligned}\hat{A}_1 = \hat{A}(\mathring{\varphi}', \mathring{\psi}', f[\mathring{\xi}'(\mathring{\delta}', \mathring{\partial}')]) = \sum_{i=1}^{i_0}\sum_{j=1}^{j_0} f(\xi'_{ij})\hat{\Psi}'_j\hat{\Phi}'_i,\\ \hat{A}_2 = \hat{A}(\mathring{\varphi}'', \mathring{\psi}'', f[\mathring{\xi}''(\mathring{\delta}'', \mathring{\partial}'')]) = \sum_{k=1}^{k_0}\sum_{l=1}^{l_0} f(\xi''_{kl})\hat{\Psi}''_l\hat{\Phi}''_k.\end{aligned} \tag{22}$$

Their difference can be written as

$$\hat{A}_1 - \hat{A}_2 = \hat{A}(\mathring{\varphi}'\circ\mathring{\varphi}'', \mathring{\psi}'\circ\mathring{\psi}'', \mathring{c}),$$

where

$$\mathring{c} = \{c_{ik,jl}\} = \{f(\xi'_{ij}) - f(\xi''_{kl})\},$$
$$i = 1, \ldots, i_0, \; j = 1, \ldots, j_0, \; k = 1, \ldots, k_0, \; l = 1, \ldots, l_0.$$

Let each of the coverings $\mathring{\delta}'$, $\mathring{\delta}''$, $\mathring{\partial}'$, $\mathring{\partial}''$ be an ε-covering and let $r(\mathring{\partial}') \leq r$, $r(\mathring{\partial}'') \leq r$. If a nonzero term in sum (22) corresponds to the system of indices (ik, jl), then we have $\delta'_i \cap \delta''_k \neq \varnothing$, $\partial'_j \cap \partial''_l \neq \varnothing$ and, consequently, we have $\rho_\Lambda(\lambda'_{ij}, \lambda''_{kl}) < 2\varepsilon$, $\rho_M(\mu'_{ij}, \mu''_{kl}) < 2\varepsilon$. Making use of the fact that the function f is uniformly continuous, we can choose ε to be sufficiently small so that when the last inequalities hold we have

$$|f(\xi'_{ij}) - f(\xi''_{kl})| < \vartheta,$$

where ϑ is an arbitrarily small positive number. Then, however, according to Lemma 4, we have

$$\|\hat{A}_1 - \hat{A}_2\|_{\hat{R}} < \vartheta r(\mathring{\partial}'\circ\mathring{\partial}'') \leqslant r^2\vartheta.$$

The convergence of the integral sums and the existence of the integral (8') then follows.

Proceeding to the limit in the inequality

$$\|\hat{A}(\mathring{\varphi}, \mathring{\psi}, f[\mathring{\xi}(\mathring{\delta}, \mathring{\partial})])\|_{\hat{R}} \leqslant rm(\mathring{\delta}, \mathring{\partial}, f[\mathring{\xi}(\mathring{\delta}, \mathring{\partial})]),$$

we obtain the estimate

$$\|Q_f\|_{\hat{R}} \leqslant r \cdot \max_{(\lambda, \mu)\in K} |f(\lambda, \mu)|.$$

This inequality shows, in particular, that functions which coincide on K have equal integrals.

The mapping $f \to Q_f$ is obviously linear and preserves involution: $Q_{\bar{f}} = (Q_f)^*$. Let us show that this mapping is multiplicative. Let f, g be functions belonging to $C(\Lambda \times M)$. Let us make use of the identity

$$\begin{aligned}\hat{A}(\mathring{\varphi}, \mathring{\psi}, f[\mathring{\xi}(\mathring{\delta}, \mathring{\partial})]) \cdot \hat{A}(\mathring{\varphi}, \mathring{\psi}, g[\mathring{\xi}(\mathring{\delta}, \mathring{\partial})]) - \\ - \hat{A}(\mathring{\varphi}, \mathring{\psi}, (fg)[\mathring{\xi}(\mathring{\delta}, \mathring{\partial})]) = \hat{A}(\mathring{\varphi}\circ\mathring{\varphi}, \mathring{\psi}\circ\mathring{\psi}, \mathring{c}),\end{aligned} \tag{23}$$

where

$$\overset{\circ}{c} = \{c_{ik,\,jl}\} = \{f(\xi_{ij})\,g(\xi_{kl}) - f(\xi_{ij})\,g(\xi_{ij})\} \tag{24}$$
$$(i,\ k = 1,\ \ldots\ i_0;\ j,\ l = 1,\ \ldots,\ j_0).$$

Let $\overset{\circ}{\delta}$, $\overset{\circ}{\partial}$ be ε-coverings. When ε is sufficiently small, the coefficients of (24) are arbitrarily small for all numbers (ik, jl) such that the sets $\delta_i \times \partial_i$ and $\delta_k \times \partial_l$ have a nonnull intersection. Therefore, applying (17) to (23) and proceeding to the limit, we obtain

$$\mathbf{Q}_f\mathbf{Q}_g = \mathbf{Q}_{fg}.$$

The multiplicative nature of the mapping $f \to \mathbf{Q}_f$ has been established.

Let $f \in C(\Lambda \times M)$ be a real function, let $\tilde{f} \in C(\Lambda \times M)$ be another function that coincides with f on the set K, and let

$$\|\tilde{f}\|_{C(\Lambda\times M)} = \max_{(\lambda,\,\mu)\in K} |f|.$$

(The function $\tilde{f}$ always exists; for example, see [17], p. 284). Let us apply estimate (18) to the sum $\hat{A}(\overset{\circ}{\varphi}, \overset{\circ}{\psi}, \tilde{f}[\overset{\circ}{\xi}(\overset{\circ}{\delta}, \overset{\circ}{\partial})])$ and then proceed to the limit. In doing this, we obtain

$$\|\mathbf{Q}_f\|_{\hat{R}} = \|\mathbf{Q}_{\tilde{f}}\|_{\hat{R}} \leqslant \max_{(\lambda,\,\mu)\in K} |f(\lambda, \mu)|.$$

This inequality can be extended to nonreal functions in the following manner:

$$\|\mathbf{Q}_f\|^2_{\hat{R}} = \|\mathbf{Q}^*_f\mathbf{Q}_f\|_{\hat{R}} = \|\mathbf{Q}_{\bar{f}f}\|_{\hat{R}} \leqslant [\max_{(\lambda,\,\mu)\in K} |f(\lambda, \mu)|]^2.$$

Thus, for any function f belonging to $C(\Lambda \times M)$, we have

$$\|\mathbf{Q}_f\|_{\hat{R}} \leqslant \max_{(\lambda,\,\mu)\in K} |f(\lambda, \mu)|.$$

Let us establish the inverse relation. To do this, we first of all note that $\mathbf{Q}_f$ is also the limit of the sums $\hat{A}(\overset{\circ}{\varphi}, \overset{\circ}{\psi}, f[\overset{\circ}{\xi}(\overset{\circ}{\bar{\delta}}, \overset{\circ}{\bar{\partial}})])$ which differ from the sums (21) in that they contain $(\lambda_{ij}, \mu_{ij}) \in \bar{\delta}_i \times \bar{\partial}_j$. Now, let $(\lambda_0, \mu_0) \in K$ be the point at which the maximum value of the function $|f|$ is attained on K. Let us construct the sum $A(\overset{\circ}{\varphi}, \overset{\circ}{\psi}, f[\overset{\circ}{\xi}(\overset{\circ}{\delta}, \overset{\circ}{\partial})])$, choosing $(\lambda_{ij}, \mu_{ij}) = (\lambda_0, \mu_0)$ for all pairs of indices (i, j) such that $(\lambda_0, \mu_0) \in \bar{\delta}_i \times \bar{\partial}_j$; for all other pairs (i, j), we choose the point (λ_{ij}, μ_{ij}) completely arbitrarily. According to Lemma 5, the following inequality holds for such sums:

$$\left\|\hat{A}\left(\overset{\circ}{\varphi}, \overset{\circ}{\psi}, f\left[\overset{\circ}{\xi}\left(\overset{\circ}{\bar{\delta}}, \overset{\circ}{\bar{\partial}}\right)\right]\right)\right\|_{\hat{R}} \geqslant |f(\lambda_0, \mu_0)|.$$

The transition to the limit yields

$$\|\mathbf{Q}_f\|_{\hat{R}} \geqslant |f(\lambda_0, \mu_0)| = \max_{(\lambda,\,\mu)\in K} |f(\lambda, \mu)|.$$

Thus, the theorem has been proved in full.

Let us assume that the function f is such that the integral

$$Q_f = \int_\Lambda \int_M f(\lambda, \mu)\,F(d\mu)\,E(d\lambda) \tag{25}$$

exists as the limit in the norm $\|\cdot\|_R$ of integral sums of the form $A(\overset{\circ}{\varphi}, \overset{\circ}{\psi}, f[\overset{\circ}{\xi}(\overset{\circ}{\delta}, \overset{\circ}{\partial})])$. It is then obvious that the integral $\mathbf{Q}_f$ also exists and that the two definitions are consistent, i.e., we have $\mathbf{Q}_f = \hat{Q}_f$. On the other hand, if the integral (25) is defined as the limiting transition for a sum of another type, then the problem of consistency becomes nontrivial. In this connection, it should be noted that, as can be easily shown, the definition of integral (25) used in [4], in which the function f is approximated by piecewise-polynomial functions, is consistent with the definition of integral (8) adopted in the present paper.

It follows from the theorem proved above that the element $\mathbf{Q}_f$ can be inverted if and only if the function f does not become zero on K and $\mathbf{Q}_f^{-1} = \mathbf{Q}_{f^{-1}}$. In other words, this means that any operator Q belonging to $\mathbf{Q}_f$ is a Noether operator and that any operator R belonging to $\mathbf{Q}_{f^{-1}}$ is a two-sided regularizer for Q.

It is obvious that the assumption of the continuity of f is a very important feature of the theory described above. The integral (8') can also be defined for a certain class of discontinuous functions, the estimate of the norm of an element of $\mathbf{Q}_f$ in terms of the uniform norm of the function f being preserved. However, the Noether property, closely associated with such estimates in the case of continuous functions, will in general no longer hold.

In conclusion, I would like to express my gratitude to M. Sh. Birman for extensive discussions of the work.

LITERATURE CITED

1. M. Sh. Birman and M. Z. Solomyak, "Stieltjes double-integral operators," Dokl. Akad. Nauk SSSR, Vol. 165, No. 6 (1965).
2. M. Sh. Birman and M. Z. Solomyak, "Stieltjes double-integral operators," in: Topics in Mathematical Physics, Vol. 1, Consultants Bureau, New York (1967).
3. M. Sh. Birman and M. Z. Solomyak, "Stieltjes double-integral operators and the factor problem," Dokl. Akad. Nauk SSSR, Vol. 171, No. 6 (1966).
4. M. Sh. Birman and M. Z. Solomyak, "Stieltjes double-integral operators. II," in: Topics in Mathematical Physics, Vol. 2, Consultants Bureau, New York (1968).
5. S. G. Mikhlin, Multidimensional Singular Integrals and Integral Equations [in Russian], Fizmatgiz (1962).
6. M. S. Agranovich, "Elliptic singular integro-differential operators," Uspekhi Matem. Nauk, Vol. 20, No. 5 (1965).
7. J. J. Cohn and L. Nirenberg, "An algebra of pseudo-differential operators," Comm. Pure and Appl. Math., Vol. 18, No. 1/2 (1965).
8. I. Ts. Gokhberg, "On the theory of multidimensional singular integral equations," Dokl. Akad. Nauk SSSR, Vol. 133, No. 6 (1960).
9. I. Ts. Gokhberg, "Some problems in the theory of multidimensional singular integral equations," Izvest. Moldavskogo Filiala Akad. Nauk SSSR, No. 10 (76) (1960).
10. R. T. Seeley, "The index of elliptic systems of singular integral operators," J. Math. Anal. and Applications, Vol. 7, No. 2 (1963).
11. R. T. Seeley, "Integro-differential operators on vector bundles," Trans. Am. Math. Soc., Vol. 117, No. 5 (1965).
12. N. Ya. Krupnik, "Multidimensional singular integral equations," Uspekhi Matem. Nauk, Vol. 20, No. 6 (1965).
13. H. O. Cordes, "The algebra of singular integral operators in R^n," J. Math. Mech., Vol. 14, No. 6 (1965).
14. M. Sh. Birman and M. Z. Solomyak, "On estimates of the singular numbers of integral operators. II," Vestnik LGU, No. 13 (1967).
15. I. Ts. Gokhberg and M. G. Krein, Introduction to the Theory of Linear Nonself-Adjoint Operators [in Russian], Izd. Nauka, Moscow (1966).
16. I. M. Gel'fand and M. A. Naimark, "Normed rings with involution and their representations," Izv. Akad. Nauk SSSR, Seriya Matem., 12 (1948).

17. P. S. Aleksandrov, Introduction to the General Theory of Sets and Functions [in Russian], Gostekhizdat (1948).

CORRECTION TO "THE INVERSE PROBLEM IN THE THEORY OF SEISMIC WAVE PROPAGATION"

A. S. Blagoveshchenskii

An error has crept into the article with the above title published in Topics in Mathematical Physics, Vol. 1, p. 55 (1967). The quantity $t(y_1)$ on p. 64 (third line from bottom) should be defined by

$$t(y_1) \equiv \int_0^{y_1} \frac{a_2(\eta_1)}{a_1(\eta_1)} d\eta_1$$

instead of the expression given in the article, namely,

$$t(y_1) \equiv \int_0^{y_1} \frac{a_1(\eta_1)}{a_2(\eta_1)} d\eta_1.$$

Despite this error, all qualitative results of the article remain in force, although some of the subsequent formulas need revision.

1. The coefficient of $\frac{1}{2}\varepsilon(t(y_1)-t)$ in formula (61) should be

$$\lambda(y_1^{\pm}) \frac{a_1(y_1^{\pm})\sqrt{a_2(y_1^{\pm})a_2(0)}}{a_2(y_1^{\pm}) \pm a_1(y_1^{\pm})} [a_2(y_1^{\pm})\sigma(y_1^{\pm}) \mp a_1(y_1^{\pm})].$$

(In the article it is of the same form, but the function a_2 and a_1 in the square brackets are interchanged.)

2. The coefficient of $\lambda(y_1^0)$ in formula (63) should be

$$\varkappa(y_1^0) \equiv \frac{a_1(y_1^0)\sqrt{a_2(y_1^0)a_2(0)}}{a_2(y_1^0) + a_1(y_1^0)} [a_2(y_1^0)\sigma(y_1^0) - a_1(y_1^0)].$$

(In the article, a_2 and a_1 in the square brackets are interchanged.)

3. The problem has been formulated correctly when $\sigma \neq \sigma_0$, where σ_0 is the positive root of the equation $2\sigma^2 + \sigma - 1 = 0$. (In the article $\sigma_0 = (1 + \sqrt{5})/2$ is the positive root of $\sigma^3 - 2\sigma - 1 = 0$.)

All subsequent formulas remain correct if the correct expression is taken for $\varkappa(y_1^0)$ as indicated above.

I would like to express my gratitude to A. A. Buzdin for pointing out the error.